Accession no.
36186698

KU-743-594

Salt Weathering Hazards

Salt Weathering Hazards

ANDREW GOUDIE and **HEATHER VILES**

School of Geography, University of Oxford, UK

LIS - LIBRARY

Date	Fund
17/09/15	Xg-Shr

Order No.

2651099

University of Chester

JOHN WILEY & SONS

Chichester · New York · Weinheim · Brisbane · Singapore · Toronto

Copyright ©1997 by John Wiley & Sons, Ltd.
Baffins Lane, Chichester,
West Sussex PO19 1UD, England

National   01243 779777
International (+44) 1243 779777
e-mail (for orders and customer service enquiries): cs-books@wiley.co.uk
Visit our Home Page on http://www.wiley.co.uk
or http://www.wiley.com

All Rights Reserved. No part of this publication may be reproduced, stored in a retrieval system, or transmitted, in any form or by any means, electronic, mechanical, photocopying, recording, scanning or otherwise, except under the terms of the Copyright, Designs and Patents Act 1988 or under the terms of a licence issued by the Copyright Licensing Agency, 90 Tottenham Court Road, London, UK W1P 9HE, without the permission in writing of the publisher.

Other Wiley Editorial Offices

John Wiley & Sons, Inc., 605 Third Avenue,
New York, NY 10158-0012, USA

WILEY-VCH Verlag GmbH, Pappelallee 3,
D-69469 Weinheim, Germany

Jacaranda Wiley Ltd, 33 Park Road, Milton,
Queensland 4064, Australia

John Wiley & Sons (Asia) Pte Ltd, 2 Clementi Loop #02-01,
Jin Xing Distripark, Singapore 129809

John Wiley & Sons (Canada) Ltd, 22 Worcester Road,
Rexdale, Ontario M9W 1L1, Canada

Library of Congress Cataloging-in-Publication Data

Goudie, Andrew.
    Salt weathering hazard / by Andrew Goudie and Heather Viles.
        p.   cm.
    Includes bibliographical references and index.
    ISBN 0-471-95842-5
    1. Salt weathering.   I. Viles, Heather A. II. Title.
QE570.G68   1997
620.1'1223—dc21   97-20299
                                                       CIP

British Library Cataloguing in Publication Data

A catalogue record for this book is available from the British Library

ISBN 0-471-95842-5

Typeset in 10/12pt Sabon from the authors' disks by Dobbie Typesetting Ltd, Tavistock, Devon
Printed & bound by Antony Rowe Ltd, Eastbourne

This book is printed on acid-free paper responsibly manufactured from sustainable forestation, for which at least two trees are planted for each one used for paper production.

Contents

Preface

As a consequence of the research that we have undertaken over many years in deserts, coasts and cities, we have become convinced that salt weathering is a major cause of both geomorphological change and damage to human-made structures. In this book we have attempted to bring together, from a diverse range of disciplines and environments, the results of research by ourselves and by others. In this connection we are greatly indebted to the following with whom we have worked over the years: Professor R. U. Cooke, Dr I. S. Evans, Dr J. C. Doornkamp, Dr Adrian Parker, Dr Peter Bush, Dr G. Evans, Dr A. W. Magee and Dr Peter Fookes. We are also grateful to Mr C. B. Jackson for a quarter of a century of assistance with experimental work. We are most grateful to Jan Magee for typing the manuscript, to Peter Hayward and Ailsa Allen for preparing the line drawings and to Martin Barfoot for preparing the plates. Dr Clifford Price very kindly commented on a first draft of Chapter 5.

Acknowledgements

We are grateful to the following for giving us permission to use previously published material in the form of figures and tables.

The Oxford University Press for Figure 3.13
The Geological Society of America for Figure 5.3 and Tables 5.3 and 5.6
John Wiley and Sons Ltd. for Tables 2.3 and 4.11, Figures 2.18, 3.14, 3.15, 4.3, 4.4, 4.5, 4.6 and 7.2
The American Meteorological Society for Table 3.7
The International Institute for Conservation of Historic and Artistic Works, for Figure 4.5
Armand Colin SA for Figure 1.6
UNESCO Publishing for Tables 1.1 and 3.6
University of Chicago Press for Figure 6.8
Elsevier Science for Figures 3.5, 5.8, 5.9, 6.19, and Table 2.2
The Geological Society of London for Figure 2.14
The American Geographical Society for Figure 3.9 and Table 3.10
Professor F. J. P. M. Kwaad for Figure 5.4
The Cambridge University Press for Figure 6.18
Drs N. and J. Lancaster for Figure 3.10
Macmillan Magazines Ltd. for Figure 5.7
The Controller of HMSO for Table 5.9 and Figure 5.10
E. Schweizerbat'sche Verlagsbuchhandlung for Figures 4.1, 4.2, 4.7, 7.1 and Tables 4.2A, 4.3B, 4.9 and 4.10
CAB International, Tables 1.6 and 1.7
United Nations University Press for Figures 2.2, 2.3 and 2.9
Dr I. B. Campbell and Dr G. G. C. Claridge for Figure 3.5
Edward Arnold for Figure 1.3
Liverpool University Press for Figure 2.8
United Nations Environment Programme for Table 1.4

Whilst every effort has been made to trace the owners of copyright material, we take this opportunity to offer our apologies to any copyright holders whose rights we have unwittingly infringed.

1 The Hazard

INTRODUCTION

Some of the world's great cultural treasures — the Pharonic temples and the Sphinx in Egypt, the rock-hewn monuments of Petra in Jordan, the baked-brick Harappan city of Mohenjo-Daro in Pakistan, the Islamic treasures of Uzbekistan, and a range of great Christian cathedrals from the Mediterranean lands — are afflicted by salt attack. The same applies to the fabric of some of the great new cities of the Middle East, including those of Bahrain, Egypt and the United Arab Emirates. The same group of processes (known to the French as *haloclastisme* and to the Germans as *Salzsprengung*) also acts in the natural landscape, providing a potent agent for geomorphological development in a range of environments that stretch from polar valleys to tropical coastlines. Salt weathering has even been proposed as a potential cause of erosion on Mars (Malin, 1974) and of spalling in the roofs of caves (Merrill, 1900). In spite of the ubiquity and severity of salt attack, it has never been the subject of a detailed monograph. This is something we seek to rectify.

That salt does indeed attack rock and building materials has been known since antiquity. Herodotus makes reference to the phenomenon in the Nile Valley, and noticed how "salt exudes from the soil to such an extent it affects even the Pyramids" (de Selincourt translation, 1972: 133).

However, it was not until the nineteenth century that salt weathering became the subject of concentrated concern and serious study. One of the prime stimuli was the need to test the resistance of building stones and to use salt "to imitate the effect of numerous winters" (Turner, 1833). A notable example of this approach is in the *Annales de Chimie et de Physique* (Vol. 38, 1828, pp.160–192), which contains a discussion of the studies of various French workers (e.g. Brard, de Thury and Vicat) on the testing of building materials using salts, notably sodium sulphate, as an analogue for frost action. Brard tested rocks with magnesium sulphate, sodium sulphate, sodium carbonate, alum, iron sulphate and muriate of soda, and found that of these sodium sulphate was "la plus énergique et la plus active". His process involved plunging the materials into a boiling and saturated solution of that salt and letting an efflorescence form.

Some nineteenth century geologists were also struck by the power of salt attack. Notable here is Hugh Miller, who described the state of his beloved Old Red Sandstone (Miller, 1841: p. 204)

Some of the sandstone beds of the system are strongly saliferous; and these, however coherent they may appear, never resist the weather until first divested of their salt. The main ichthyolite bed on the northern shore of the Moray Firth is overlaid by a thick deposit of a finely-tinted yellow sandstone of this character, which, unlike most sandstones of a mouldering quality, resists the frosts and storms of winter, and wastes only when the weather becomes warm and dry. A few days of sunshine affect it more than whole months of high winds and showers. The heat crystallises at the surface the salt which it contains; the crystals, acting as wedges, throw off minute particles of the stone; and thus, mechanically at least, the degrading process is the same as that to which sandstones of a different but equally inferior quality are exposed during severe frosts. In the course of years, however, this sandstone, when employed in building, loses its salt; crust after crust is formed on the surface, and either forced off by the crystals underneath, or washed away by the rains; and then the stone ceases to waste, and gathers on its weathered inequalities a protecting mantle of lichens.

Major developments in the study of salt weathering came as geomorphologists travelled to desert environments from the end of the nineteenth century onwards. Particular attention needs to be drawn to the influential work of the German geomorphologist, J. Walther, whose work in Egypt, and in particular his observations of decayed monuments, made him a firm convert to the power of "exsudation" (Walther, 1912). Walther in turn had a powerful influence on the American geomorphologist Hobbs (1917: 201–202), who likened salt crystallisation within a rock's capillary fissures to water freezing within a pipe. The belief of German geomorphologists in the power of salt weathering continued into the 1930s, when Mortensen (1933) produced his major essay on the salt hydration mechanism.

A renewed spasm of interest in salt weathering arose in a whole series of different centres from the 1950s onwards. French climatic geomorphologists such as Birot and Tricart were one active group, and they undertook important experiments and field observations. The accelerating tempo of polar science, particularly in the Dry Valleys of Antarctica, led to a large body of observations by New Zealand and American scientists (these are reviewed by Evans, 1970). Two other stimuli to the growth of salt weathering studies were applied geomorphological studies by British geomorphologists working with engineering geologists in the Middle East (see Cooke et al, 1982) and a burgeoning concern with the effects of air pollution (including acid deposition) on building stones in Europe and elsewhere (see, for example, Pühringer, 1983). Indeed, as the next section shows, there are a variety of grounds for fearing that salt weathering is becoming a more serious hazard in many parts of the world.

On a global basis the presence of salt in the environment is a widespread phenomenon, especially in drylands and some coastal environments. Salts are also accumulating in urban environments because of air pollution which, for example, produces sulphates and nitrates that cause building stone decay. Table 1.1 provides an estimate of salt-affected soils by continent and subcontinent and suggests that the total area amounts to 952 million hectares.

Table 1.1. Extent of salt-affected* soils by continent and subcontinent (after Szabolcs, 1979). Reproduced by permission of Unesco

Region	Area affected (10^6 ha)
North America	15.7
Mexico and Central America	2.0
South America	129.2
Africa	80.5
South Asia	84.8
North and Central Asia	211.7
South-east Asia	20.0
Australia	357.3
Europe	50.8
Total	952.0

*Note: "salt-affected" includes both saline soils and alkaline soils.

Wellman and Wilson (1965: 1097) argued that salt weathering would be important in two main types of situation: regions where salts are being concentrated and regions where the concentration of salts is kept high by a rapid rate of supply

The first is represented by arid regions and by micro-environments that are protected from rain, such as under ledges, on the insides of caves and parts of buildings. Salts are provided by air-borne salt particles, from rain and snow which everywhere contains small quantities of salts, from chemical degradation of the rocks themselves, and from ground water percolating to the surface evaporating and depositing salts. The second is represented by the sea coast in most parts of the world. Salts are provided at a rapid rate by splashing of waves and from sea spray. The rocks most affected are those that frequently become wet and then dry, that is, in general, those in the zone immediately above the reach of waves at high water.

Like many geomorphological processes, salt weathering can pose a hazard if it affects human artefacts and livelihoods. Unlike more dramatic processes, such as landslides, avalanches and coastal erosion, salt weathering is a slow, creeping hazard. Nevertheless it can, as we shall see throughout this book, have serious economic effects as well as being capable of irreparably damaging items of our cultural heritage. Like most environmental hazards, salt weathering is at least partially produced by human activity, through placing vulnerable materials in salt-affected environments, or through increasing the presence and destructive capabilities of salts, or a combination of both.

IS THE SALT HAZARD INCREASING?

Although, as we have already noted, salinity is a normal occurrence in a wide range of environments, in some parts of the world the extent and severity of salinity is increasing as a result of a whole range of human activities (Table 1.2).

Of these activities, the most important is the spread of irrigation (Table 1.3). This causes increased levels of salinity in a variety of ways (see Rhoades, 1990). Firstly, the application of irrigation water to the soil leads to a rise in the water-table so that it may become near enough to the ground surface for capillary rise and subsequent evaporative concentration to take place. When groundwater comes within 3 m of the surface in clay soils, and even less for silty and sandy soils, capillary forces bring moisture to the surface where evaporation occurs. There is plenty of evidence that irrigation does indeed lead

Table 1.2. Causes of aggravated salt attack

Irrigation
   Rise in groundwater
   Evaporation of water from fields
   Evaporation of water from reservoirs
   Waterlogging produced by seepage losses
Sea water incursion
   Caused by over-pumping
   Reduced freshwater recharge
Vegetation clearance
Inter-basin water transfers
   Mineralisation of lake waters
   Deflation of salts from desiccated lakes
Urbanisation
   Water importation, faulty drains, etc.
Atmospheric pollution
Salting of roads, etc.
Changes in the internal microclimates of buildings

Table 1.3. Estimates of global irrigated land

Year	Irrigated area (10^6 ha)
1900	44–48
1930	80
1950	94
1955	120
1960	168
1980	211
1990 (estimate)	240

Sources: FAO and other data in Gleick (1993: table E.2) and Heathcote (1983: table 13.2).

to rapid and substantial rises in the position of the water-table, as shown by the data in Table 1.4. Rates typically range between 0.2 and 3 m per year.

Secondly, many irrigation schemes, being in areas of high temperatures and high rates of evaporation, suffer from the fact that the water applied over the soil surface is readily concentrated in terms of any dissolved salts it may contain. This is especially true for crops with a high water demand (e.g. rice) or in areas where, for one reason or another, farmers are profligate in their application of water (Figure 1.1).

Thirdly, the construction of large dams and barrages creates large water bodies from which further evaporation can take place, once again leading to the concentration of dissolved solutes.

Fourthly, notably in sandy soils with high permeability, water seeps both laterally and downwards from irrigation canals so that waterlogging may occur. Many irrigation canals are not lined, with the consequence that substantial water losses can result (Figure 1.2).

These four processes are of considerable significance given the rapid expansion in irrigation that has taken place in recent decades. On a global basis, the calculations of Rozanov et al (1991: p. 120) make grim reading. They estimate that "from 1700 to 1884 the global area of irrigated land increased from 50,000 to 2,200,000 km², while at the same time some 500,000 km² were abandoned as a result of secondary salinization". During the 1950s the irrigated area was increasing at over 4% annually, though the figure has now dropped to only about 1%. The amount of salinised irrigated land varies from area to area (Table 1.5), but in general ranges between 10 and 50% of the total. It needs to be noted, however, that there is a considerable range in these values according to the source of the data (compare Table 1.6) and this may in part reflect differences in the definition of the terms

Table 1.4. Increase in level of water-tables due to irrigation

Irrigation project	Country	Water-table height (m)	
		Original depth	Rise per year
Nubariya	Egypt	15–20	2.0–3.0
Beni Amir	Morocco	15–30	1.5–3.0
Murray–Darling	Australia	30–40	0.5–1.5
Amibara	Ethiopia	10–15	1.0
Xinjang Farm 29	China	5–10	0.3–0.5
Bhatinda	India	15	0.6
SCARP 1	Pakistan	40–50	0.4
SCARP 6	Pakistan	10–15	0.2–0.4

Source: Tolba and El-Kholy, 1992, table 5: 94. Reproduced by permission of The United Nations Environment Programme.

Figure 1.1. The inevitable consequences of the spread of irrigation are that water-tables rise, waterlogging occurs and salts accumulate in soils. This irrigated field is located on the northern fringes of the Sahara in Morocco on the south side of the High Atlas (photograph by A. S. Goudie)

Figure 1.2. Over-irrigation in the Indus Plain, Pakistan, has caused waterlogging and the accumulation of salt — "a satanic mockery of snow". The spread of such conditions is a common phenomenon in the world's arid lands (photograph by A. S. Goudie)

Table 1.5. Salinisation of irrigated cropland

Country	Percentage of irrigated lands affected by salinisation
Algeria	10–15
Australia	15–20
China	15
Colombia	20
Cyprus	25
Egypt	30–40
Greece	7
India	27
Iran	<30
Iraq	50
Israel	13
Jordan	16
Pakistan	<40
Peru	12
Portugal	10–15
Senegal	10–15
Sri Lanka	13
Spain	10–15
Sudan	<20
Syria	30–35
USA	20–25

Source: FAO data as summarised in World Resources (1987, 1988): table 19.3.

Table 1.6. Global estimate of secondary salinisation in the world's irrigated lands

Country	Cropped area (Mha)	Irrigated area (Mha)	Share of irrigated to cropped area (%)	Salt-affected land in irrigated area (%)	Share of salt-affected to irrigated land (%)
China	96.97	44.83	46.2	6.70	15.0
India	168.99	42.10	24.9	7.00	16.6
CIS	232.57	20.48	8.8	3.70	18.1
USA	189.91	18.10	9.5	4.16	23.0
Pakistan	20.76	16.08	77.5	4.22	26.2
Iran	14.83	5.74	38.7	1.72	30.0
Thailand	20.05	4.00	19.9	0.40	10.0
Egypt	2.69	2.69	100.0	0.88	33.0
Australia	47.11	1.83	3.9	0.16	8.7
Argentina	35.75	1.72	4.8	0.58	33.7
South Africa	13.17	1.13	8.6	0.10	8.9
Subtotal	842.80	158.70	18.8	29.62	20.0
World	1473.70	227.11	15.4	45.4	20.0

Source: Ghassemi et al, 1995, table 18. Reproduced by permission of CAB International and the University of New South Wales Press.

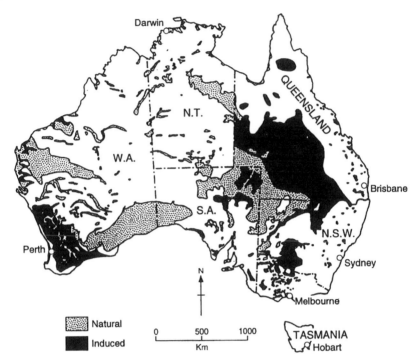

Figure 1.3. Natural and human-induced salinisation in Australia. Modified after Australian Standing Committee on Soil Conservation, Canberra (1982). Reproduced from Williams and Balling (1996) by permission of Edward Arnold/Hodder & Stoughton Educational

"waterlogging" and "salinisation". Figure 1.3 is a map of Australia showing the areas that are believed to be the result of human-induced salinisation.

Trends in the salinity of rivers brought about by these processes have been identified in some areas. The most impressive data come from Western Australia (Table 1.7) where the record, in some cases, dates back to 1940. The rate of increase in stream salinity has been rapid, with values since 1965 ranging between 11 and 117 mg l^{-1} yr^{-1}.

A good example of the effects of rising groundwater levels caused by irrigation is provided by the weathering of important Islamic archaeological sites in Uzbekistan (Cooke, 1994), in cities such as Khiva, Bukhara and Samarkand. Irrigation has increased rapidly in the region since the Russian conquest in the late nineteenth century, and especially since 1945, to sustain cotton production.

A second prime cause of the spread of saline conditions is the incursion of sea water brought about by the over-pumping of groundwater. Ocean water displaces less saline groundwater through a mechanism called the Ghyben–Herzberg principle. The problem presents itself on the coastal plain of Israel,

Table 1.7. Stream salinity of major rivers in western Australia

Catchment area	Period of record	Area cleared (%)	Average stream salinity over last 5 years of record (mg l^{-1})	Rate of stream salinity increase over period of record (mg l^{-1} yr^{-1})	Rate of stream salinity increase since 1965 (mg l^{-1} yr^{-1})
Denmark R	1960–86	17	890	25	26
Kent R	1956–86	40	1870	52	58
Frankland R	1940–86	35	2192	44	74
Warren R	1940–86	36	870	12	15
Perup R*	1961–86	19	3410	132	117
Wilgarup R*	1961–86	33	863	20	14
Blackwood R	1956–86	85	2192	52	58
Capel R	1959–76	50	423	15	14
Preston R	1955–75	50	354	8	11
Thomson R	1957–85	45	534	18	17
Collie R	1940–86	24	730	11	24
Murray R	1939–86	75	2792	39	93
Williams R†	1966–86	90	2425	95	95
Hotham R†	1966–86	85	3711	89	89
Woorollo Bk	1965–86	50	2092	44	39
Brockman R	1963–86	65	2040	76	72
Helena R	1966–85	10	1257	48	48

*Tributaries of the Warren River.
†Tributaries of the Murray River.
Source: Ghassemi et al, 1995, table 2.6. Reproduced by permission of CAB International and the University of New South Wales Press.

in parts of California, on the island of Bahrain, and in some of the coastal aquifers of the United Arab Emirates. A comparable situation has also arisen in the Nile Delta, though here the cause is not necessarily groundwater over-pumping, but may be due to changes in water levels and freshwater recharge caused by the construction of the Aswan High Dam. Figure 1.4 shows the way in which chloride concentrations have increased and spread in the Llobregat Delta area of eastern Spain because of the incursion of sea water.

A third prime cause of soil salinity extension is vegetation clearance. The removal of native forest vegetation allows a greater penetration of rainfall into deeper soil layers which causes groundwater levels to rise, thereby creating conditions for the seepage of sometimes saline water into low-lying areas. This is a particularly serious problem in the wheatbelt of Western Australia and in some of the prairie areas of North America. In the case of the former area it is the clearance of *Eucalyptus* forest that has led to the increased rate of groundwater recharge and to the spreading salinity of streams and bottom-lands. Salt "scalds" have developed. The speed and extent of groundwater rise

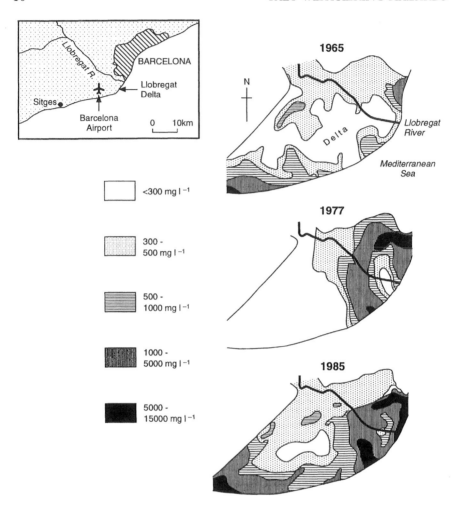

Figure 1.4. Changes in the chloride concentration of the Llobregat Delta aquifer, Barcelona, Spain as a result of sea water incursion caused by the over-pumping of groundwater. Modified after Custodio et al, 1986

following such forest clearance is shown in Figure 1.5. Until late 1976 both the Wights and Salmon catchments were forested, then the Wights was cleared. Before 1976 both catchments showed a similar pattern of groundwater fluctuation, but after that date there was a marked divergence of 5.7 m (Peck, 1983). The process can be reversed by afforestation (Bari and Schofield, 1992).

A further reason for increases in levels of salinity is the changing state of water bodies caused by inter-basin water transfers. The most famous example of this is the shrinkage of the Aral Sea, the increase in its mineralisation, and

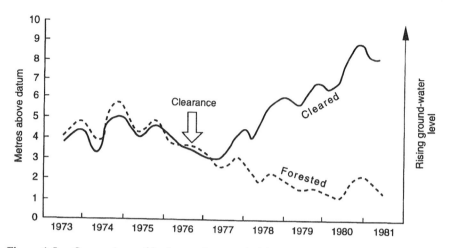

Figure 1.5. Comparison of hydrographs recorded from the boreholes in Wights (solid line) and Salmon (broken line) catchments in Western Australia. Both catchments were forested until late in 1976 when Wights was cleared. Modified after Peck (1983: figure 1)

the deflation of saline materials from its surface and their subsequent deposition downwind. Although estimates of the amount of salty material being deflated from the desiccating floor of the sea vary, they agree that some tens of millions of tonnes of salt are being translocated by dust storms (Figure 1.6) each year. The sea itself has had its mineral content increased more than threefold since 1960.

Urbanisation can lead to changes in groundwater conditions that can aggravate salt attack. In some large desert cities the importation of water, its usage, wastage and leakage, can produce the ingredients to feed this phenomenon. This has, for example, been identified as a problem in Cairo and its immediate environs (Hawass, 1993). The very rapid expansion of Cairo's population has outstripped the development of an adequate municipal infrastructure. In particular, leakage losses from water pipes and sewers have led to a substantial rise in the groundwater level and have subjected many buildings to attack by sulphate- and chloride-rich water. The medieval Qalawun Complex is one example.

There are other sites in Egypt where urbanisation has been identified as a major cause of accelerated salt weathering of important monuments. Smith (1986: 510), for example, described the damage to tombs and monuments in Thebes and Luxor in the south of the country.

> While the structures are now protected from outright destructive exploitation, more subtle factors are responsible for their decay. Many of the recent problems may be traced either directly or indirectly to the Aswan High Dam. Water afforded by the dam for irrigation purposes has been used indiscriminately by

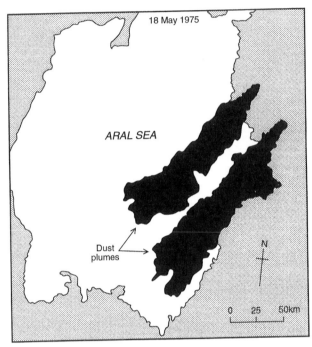

Figure 1.6. Dust plumes caused by the deflation of salty sediments from the drying floor of the Aral Sea as revealed by a satellite image (153/Meteor-Priroda, 18 May 1975). Modified after Mainguet (1995: Figure 4) by permission of Armand Colin

farmers whose fields adjoin the ancient monument sites. Excessive moisture from over-watering and inadequate drainage now seeps into the foundation material, resulting in severe damage.

In the context of urbanisation it also needs to be appreciated that recent decades have seen a marked increase in the exposure of buildings to salt hazard because of the rapid expansion of urban settlements in some potentially aggressive areas. This is, for example, the case with some of the Gulf States of the Middle East. The city of Abu Dhabi had a population of around 8000 in 1960. By 1984 it had reached 243 000, a 30-fold increase. The city of Dubai had a population of around 45 000 in 1960. By 1984 this had reached 266 000, representing an almost six-fold increase.

Urbanisation can also cause a rise in groundwater levels by affecting the amount of moisture lost by evapotranspiration. Many elements of urbanisation, and in particular the spread of impermeable surfaces (roads, buildings, car parks, etc.) interrupt the soil evaporation process so that groundwater levels in *sabkha* (salt plain) areas along the coast of the Arabian Gulf rise at a rate of 40 cm yr^{-1} until a new equilibrium condition is attained; the total rise from this cause may be 1–2 m (Shehata and Lotfi, 1993). This can require the construction of horizontal drains.

Changes within the buildings themselves can lead to an acceleration in salt weathering. A clear example of this is the effect that modern heating systems have on air temperature and humidity and through them on salt action. Arnold and Zehnder (1988) discuss this in connection with the weathering environment in Swiss churches, where intense central heating can cause low internal relative humidities and the deleterious crystallisation of hygroscopic salts. Likewise, Laue et al (1996) found in the crypt of St Maria im Kapitol in Cologne, Germany that salts dissolve in the summer months, but as soon as the central heating comes on in October the salts effloresce and they develop progressively until the spring, causing catastrophic stone decay.

Rock salt (sodium chloride) has been used in increasing quantities since the Second World War for minimising the dangers to motorists and pedestrians from icy roads and pavements (Figure 1.7). With the rise in the number of vehicles there has been a tendency throughout Europe and North America for a corresponding increase in the use of salt for de-icing purposes (see, for example, Howard and Beck, 1993). Data for the UK between 1960 and 1991 are shown in Figure 1.8 and demonstrate that while there is considerable inter-

Figure 1.7. Development of a salt efflorescence with associated alveole and tafoni development on a limestone wall in Woodstock, Oxfordshire, UK. It is possible that salt spray from de-icing salts applied to the main road in the foreground may be responsible. Rock flour has accumulated at the base of the wall (photograph by A. S. Goudie)

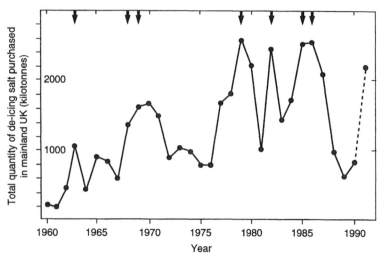

Figure 1.8. Estimates of the total quantity of de-icing salt purchased annually in mainland Britain during the period 1960–1991. Arrows highlight severe winters. Data provided by ICI in Dobson (1991: figure 1.1)

annual variability (related to weather severity), there has also been a general upward trend so that in the mid-1970s to end 1980s more than two million tonnes of de-icing salt were being purchased each year (Dobson, 1991). Data on de-icing salt application in North America are given by Scott and Wylie (1980). They indicate that the total use of de-icing salts in the USA increased at a nearly exponential rate between the 1940s and the 1970s, increasing from about 200 000 metric tonnes in 1940 to approximately 9 000 000 metric tonnes in 1970. This represented a doubling time of about five years. There is now an extensive literature on the effects of de-icing agents on concrete and, in particular, on bridge decks (see, for example, Litvan, 1975; 1976).

Many examples of the decay of modern bridges as a result of salt attack are given in Mallett (1995). They include the Tay road bridge in Scotland, the Tees Viaduct in north-east England and various bridges in Saudia Arabia. De-icing salts are implicated in the British examples. Indeed, chloride contamination due to the use of road salt is considered to be the most serious cause of the corrosion problem affecting at least 60 000 road bridges in Britain (Mallett, 1995: 145). In North America the problem is believed to be substantially more serious than in the UK (Figure 1.9), with an estimated 19% of bridges requiring replacement or repair at a cost of more than $23 000 million at 1977 prices (Vassie, 1984: 713).

Specific examples of accelerated salt weathering, produced by the sorts of changes just mentioned, include the decay of the great Harappan

Figure 1.9. Severe weathering taking place on a flyover support on the road system in central Toronto, Ontario, Canada. It is likely that damage of this sort is caused in part by de-icing salts. The iron reinforcements are being corroded, their volume is increasing and concrete spalling is taking place (photograph by A. S. Goudie)

archaeological site at Mohenjo Daro in Pakistan (see Chapter 2) and the disintegration of Islamic buildings in Uzbekistan, Central Asia (see Chapter 2).

With regard to the future, global changes in climate and sea level could have implications for salinity conditions and the magnitude of the salt weathering hazard. For example, increasing drought risk in Mediterranean Europe could lead to a substantial increase in salt-prone areas (Imeson and Emmer, 1992; Szabolcs, 1994), while higher sea levels in susceptible geomorphological locations (coastal deltas, the sabkhas of the Arabian Gulf) could change the position of the all-important salt–freshwater interface and the height of the water-table. It is also possible, as Hall and Walton (1992) have argued, that salt weathering could be enhanced in polar regions as freeze–thaw regimes alter. Indeed, looking at Table 1.2 it is clear that global warming and its allied environmental changes could influence many of the causes of aggravated salt attack either directly or indirectly.

A major cause of salt weathering, especially in urban areas, is atmospheric pollution. Gaseous pollutants such as sulphur dioxide, nitrogen oxides and hydrochloric acid all contribute to salt weathering, and particulates (such as carbonaceous spheres produced by fossil fuel combustion) and oxidants (such as ozone produced by photochemical reactions) may act as important catalysts. Sulphur dioxide, for example, reacts with dust and other particles and reactive materials such as many building stones to form gypsum (Figure 1.10). An important part of this process is the oxidation of sulphur dioxide either in the gas phase or in moisture films on building stones. This oxidation is assisted by atmospheric oxidants, such as ozone and hydrogen peroxide, as well as by catalysts such as soot and smoke.

The levels of various key atmospheric pollutants have changed greatly over recent years, producing an additional dynamism to the salt weathering hazard. In many developed countries, for example, changes in fuel usage and economic activity have resulted in a steady decline in sulphur dioxide concentrations from their peak in the 1970s (Mylona, 1996). In these same countries, however, levels of nitrogen oxides have tended to increase over the same period as a result of burgeoning traffic levels (as petrol- and diesel-powered vehicles are a major source of nitrogen oxides) producing higher levels of

Figure 1.10. The balustrade wall of the Ashmolean Museum, Oxford constructed of Portland Stone, shows the clear contrast between white, water-washed areas and sheltered black, soot and gypsum-encrusted areas. During recent building works at the museum this balustrade wall has been cleaned (photograph by H. A. Viles)

nitrate salts in the environment. In many rapidly industrialising countries sulphates and nitrates are both increasing, as domestic, industrial and transport uses of fossil fuels boom, producing a highly corrosive urban environment.

Hydrochloric acid is an important pollutant source of chloride ions, adding to that of sea spray in many areas. Coal combustion is the major non-natural source of hydrochloric acid. British coals, for example, contain 0.23% chloride on average, and of this 94% is emitted as hydrochloric acid in stack gases (Hutchinson et al, 1992). In Western Europe hydrochloric acid contributes <2% of the total acidity, with NO_x contributing 3% and 68% attributable to sulphur dioxide. In the UK, hydrochloric acid contributes 4%, with 25% coming from nitrogen oxides and 71% from sulphur dioxide. Hydrochloric acid is a reactive gas, quickly removed from the atmosphere close to its source by contact with cloud water and other surfaces.

Polluted areas near coasts can suffer from a particularly potent cocktail of gases and particulates producing a wide range of damaging salts under favourable atmospheric conditions, such as experienced by Berkeley, California, where fogs can be particularly corrosive.

CONCLUSIONS AND CONCEPTUAL FRAMEWORK

Salt weathering is perceived by many to be an important, often hazardous, process in a wide range of areas. Urban, coastal and dryland environments are particularly prone to its impacts. Important sources of salts include airborne material (whether from sea spray, air pollution or the reworking of other deposits) and groundwater. Environmental conditions which favour evapo-transpiration increase the risk of damage from such salts. Salt weathering affects buildings, engineering structures, rock outcrops and minerals within the soil profile and there is compelling evidence that its influence is increasing and will continue to do so as human impacts continue to affect both the local environment and the global climate. There are still many gaps in our knowledge of salt weathering processes, despite an urgent need to understand and remedy their effects, but the powerful combination of field observations, laboratory experimentation and modelling is continually refining our appreciation of, and ability to control, the salt hazard.

From the preceding review of the salt weathering hazard it is clear that several linked components need to be understood before the actions of salt on natural materials, building materials and other artefacts can be comprehended and appropriate remedial action taken. As shown in Figure 1.10 salt weathering requires salts, sources of moisture, suitable environmental conditions and susceptible material. It results in the formation of new products and the creation of small-scale topographic features (which produce unsightly blemishes on affected buildings and may also contribute to landform development in natural settings).

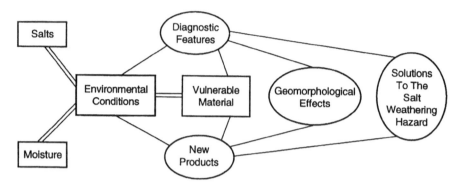

Figure 1.11. Conceptual framework for understanding the salt weathering hazard

The rest of this book investigates some of the major facets of salt weathering, following the framework provided by Figure 1.11. In so doing, we draw upon a vast and diverse range of published material as well as our own field and laboratory observations. Chapter 2 presents a range of case studies from around the world to stress the importance and diversity of the salt weathering hazard. In Chapter 3 we discuss the major sources of salt and moisture and the environmental conditions which favour salt weathering. Chapter 4 considers the important features of materials which make them susceptible to salt attack, and then reviews the progress made by experimental weathering studies. Such experimental studies have been particularly important in understanding salt weathering and also form the basis for many investigations of the utility of a range of compounds to reduce the salt weathering hazard. Chapter 5 builds on the previous chapters to provide a state of the art review of how salts produce weathering, drawing on field, laboratory and modelling studies. The geomorphological ramifications of salt weathering are discussed in Chapter 6, and Chapter 7 looks at a range of possible solutions to the salt weathering hazard.

2 Case Studies

INTRODUCTION

Salt weathering is a major hazard for buildings and other engineering structures in a range of environments. In this section we discuss a suite of case studies that relate to a diverse range of environments, types and ages of structure. In so doing we aim to demonstrate that salt attack strikes both at some of the world's great cultural monuments and at some of its most modern urban areas. Examples of the former are listed in Table 2.1.

MOHENJO-DARO, PAKISTAN

In the arid Indus valley of Pakistan and neighbouring areas, a remarkable civilisation flourished from about 2300 to 1750 BC (Allchin and Allchin, 1968: 140). Some of the settlements of this civilisation have been excavated over the last 70 years. The largest and most important of these Harappan or Indus civilisation sites is Mohenjo-Daro (Moenjodaro) in the Larkana district of Sind. This has been excavated by numerous workers, including Marshall, Mackay, Wheeler and Dales, since its discovery in 1922.

It is situated on the Indus plain, where the general height is about 47 m above sea level. The excavated parts of the ruins lie almost entirely between the plain level and the 49 m contour. However, the highest part of the site, the Buddhist Stupa Mound, rises to more than 61 m. The median flood level of the Indus is about 47 m and the high flood level about 50 m. The groundwater level fluctuates seasonally by about 2.4 m, from 1.5 to 3.9 m below plain level. Thus the highest parts of the site lie about 16 m above the groundwater level. The groundwater itself has a moderate salinity, with the total dissolved solids ranging from 416 to 3419 ppm (Unesco, 1964).

The Master Plan for Mohenjo-Daro (1972) states (p. 12): "It is a paradox of facts that the ruins of Moenjodaro buried beneath the accumulations of five thousand years remained in an excellent state of preservation. But as soon as they were exposed from oblivion to the incredible gaze of the 20th century, they were overtaken by the plague of water logging and the leprosy of salinity". Likewise, Unesco (1964: 17) report "the brick paths laid for the visitors are suffering heavy and rapid damage by exfoliation and disintegration

Table 2.1. Examples of archaeological and architectural heritage sites being damaged by salt attack

Location	Reference
Temples of Karnak, Egypt	Bromblet (1993)
Temples of Luxor and Thebes, Egypt	Smith (1986)
Sphinx, Egypt	Hawass (1993)
Mohenjo Daro, Pakistan	Goudie (1977), van Lohuizen-de Leeuw (1974)
Medieval mosques, Ras al Khaimah (UAE)	This book
Mosques, Anatolia, Turkey	Tuncoku et al (1993)
Islamic buildings, Uzbekistan	Cooke (1994)
Petra, Jordan	Albouy et al (1993), Fitzner and Heinrichs (1991; 1994)
Venetian fortress, Corfu	Moropoulou et al (1993)
Marble statues in Delos and Naxos	Beloyannis and Dascalakis (1990)
Church of Notra Dame La Grande (12th century), Poitiers, France	Hammecker and Jeannette (1988)
Saint Mark's Basilica, Venice, Italy	Fassina (1988)
Seville Cathedral, Spain	Alcalde and Martin (1988)
Lincoln Cathedral, UK	Butlin et al (1988)
Wallpaintings, plaster and stucco in Swiss churches	Arnold and Zehnder (1988), Zehnder (1996)
Castle, Rhodes, Greece	Theolakis and Moropoulou (1988)
Monuments in Apulia, s. Italy including Bari Cathedral	Zezza and Macrì (1995)
Trinity College, Dublin, Ireland	O'Brien et al (1995)
Cathedral of Toledo, Spain	La Iglesia et al (1994)
S. Maria dei Miracoli Church, Venice, Italy	Fassina et al (1996)
Crypt of St Maria im Kapitol, Cologne, Germany	Laue et al (1996)
St Mary's Cathedral and Fruedenstein Castel, Freiberg, Saxony, Germany	Klemm and Siedel (1996)
Hospital de Santo António, Oporto and Edifício do Largo do Paço, Braga, Portugal	Begonha et al (1996)
Don Cathedral, Riga, Latvia	Vitina et al (1996)
Almeria Cathedral, Spain	Villegas Sánchez et al (1996)
Malaga Cathedral, Spain	Carretero and Galan (1996)
Palau de la Generalitat, Barcelona, Spain	Garcia-Vallès et al (1996)

of the bricks. Many details of the ancient architecture have been lost within only a few years by the same cause". This disintegration has also been noted by archaeologists.

Sir John Marshall himself, who excavated the site in the 1920s, wrote (Marshall, 1973 edn: 1)

The salts also which permeate the soil of Sind have hastened the decay of the site. With the slightest moisture in the air, these salts crystallize on any exposed surface of the ancient brickwork, causing it to disintegrate and flake away, and eventually reducing it to powder. So rapid is their action that within a few hours after a single shower of rain newly excavated buildings take on a mantle of white rime like freshly fallen snow. The desolation that thus distinguishes this group of mounds is shared by the plain immediately around them, which for the most part is also white with salt....

The American archaeologist, Mackay, also noted the rapid rate of brick disintegration that could occur (Mackay, 1938, quoted in Jansen, 1996)

... Those who have seen the rapid crumbling of burnt brick under the action of salt, even after quite a small shower of rain, will readily realise what great damage water and damp can do to buildings so impregnated with salt as those of Mohenjo-Daro. I have seen a brick crumble to powder the first five days after rain, and the tops of walls are converted into layers of dust within a week.

Van Lohuizen-de Leeuw (1973) comments that "the astounding remains of Moenjo-Daro... are for the greater part in a state of utter disintegration and decay and are rapidly approaching the point of total destruction" (p. 1). She continues, "Moenjo-Daro the City of Dead, is indeed rapidly decaying and approaching its own death".

Thus it is clear that the disintegration of the burnt bricks has accelerated since the mid-1920s, before which they were in a state of relatively good preservation. In that time many have changed from regularly sized $(0.285 \text{ m} \times 0.135 \text{ m} \times 0.055 \text{ m})$ bricks into amorphous piles of dust (Figure 2.1). However, bricks affected by this disintegration are being replaced continuously by comparable burnt bricks of the normal size or by "English" bricks $(0.23 \text{ m} \times 0.125 \text{ m} \times 0.063 \text{ m})$. Some of these bricks are also being rapidly destroyed. In December 1975 brick paths and some walks laid as recently as 1972 and June 1974 were observed to have disintegrated. Artificial building materials were not the only ones to be thus affected. Limestone aggregate used for the sub-grade of roads (built in 1971–1972) and for ornamental effects in front of the Rest House (built 1964) were also found to be locally disintegrated, often to powder. In general, therefore, it is clear that in favourable situations, such as at ground level and on the lower parts of exposed walls, the disintegration of natural and artificial materials takes place with great efficiency, a time span of 2–12 years often being sufficient for complete breakdown to occur (Goudie, 1977).

The disintegration is associated with the development of a white efflorescence on brick and stone surfaces. Disintegration is minimal where there is no efflorescence. The disintegration is caused by the hydration and crystallisation of various salts. The growth of the crystals has been noted as occurring daily by the Unesco team (Unesco, 1964: 46): "At night and very early in the morning, when it was cold and the humidity of the air was relatively high, the floors and the lower parts of the walls of the various

Figure 2.1. Salinisation and disintegration of bricks by sodium sulphate attack at the Harappan archaeological site of Mohenjo-Daro, Sind, Pakistan (photograph by A. S. Goudie)

buildings became damp. With rising temperatures and lowering humidity of the air, the evaporation exceeded the flow of capillary water. As a result the surfaces of the floors and walls dried out and in the morning one could observe the rapid growth of needle like crystals of salts. A growth of bunches of needle like crystals repeated every morning anew".

The development of the salt efflorescence results from several different processes. Over much of Sind the development of modern irrigation systems has led to an increase in the height of groundwater levels, to waterlogging and to consequent salinity.

At Mohenjo-Daro, when excavations started in 1922, the water-table was about 7.6 m below the surface, whereas now it averages only about 2 m. This phenomenon is especially serious in areas of extensive rice cultivation such as the Larkana district. It is clear that some of the salt is derived from high groundwater levels and from waterlogging, and that the salt efflorescence develops at the upper limit of the capillary fringe. However, because the height of the capillary fringe above the water-table in alluvial silts is usually of the order of 3 m [Jenkins (1932: 112) gives a figure of 1.2–1.8 m for Sind], it is likely that this mechanism is only effective at lower levels. Higher up the mound of Mohenjo-Daro, which, as already noted, rises to 16 m above the groundwater level, other sources of salt are probably more important. These include the initial salt content of the bricks (themselves frequently manufactured in areas of moist and, therefore, potentially saline alluvium), and inputs from the atmosphere in the form of rain and dust (though analytical data on these are not available). Around Mohenjo-Daro large expanses of fields are covered by white efflorescence, and some of this salt may be transported by aeolian processes to accumulate among the walls of the mound. It is unlikely that the proposed tubewell scheme to reduce groundwater levels (Master Plan, 1972) will have much effect in terms of the salts derived directly from the bricks and debris of the mound or by aerosolic inputs.

Although the presence of large quantities of any salts might in itself be sufficient to cause disintegration, it appears likely that the extreme rapidity of this process in the Sind plain is caused by the nature of the salts comprising the efflorescence. Even though the efflorescence is composed of more than one salt, the predominant component is sodium sulphate, with some other sulphates (including calcium sulphate, magnesium sulphate and potassium sulphate) also present.

X-Ray diffraction analysis was carried out on selected samples at Mohenjo-Daro to obtain a measure of the minerals present. The minerals identified include quartz, mica, chlorite, aphthitalite, burkeite, sodium carbonate and thenardite. The three most important minerals identified are thenardite, aphthitalite $[(K, Na)_3 Na (SO_4)_2]$ and burkeite $[Na_6(CO_3)(SO_4)_2]$. The quartz, mica and chlorites are impurities introduced into the efflorescence by wind-transported dust and by the disintegration of bricks and mortar.

SUEZ CITY, EGYPT AND DUBAI, UAE

In the 1970s it became evident in the planning of new urban developments in the Middle East that salt weathering was a potential threat for substantial capital investments. To that end, groups of British geomorphologists were involved in site assessments at Suez City, Egypt and at Dubai in the United Arab Emirates. As in Bahrain, analyses were undertaken of the groundwater depth and of levels of salinity (expressed as conductivity). In both cities there were extensive areas where groundwater levels were high and salt contents elevated.

At Suez City salinity values locally reached $388\,000\,\mu$mhos cm^{-1} ($1\,\mu$mhos cm$^{-1}=0.1$ mS m^{-1}) (Jones, 1980: 66) and in places the water-table was less than 1 m below the surface. It was estimated that salt weathering due to the capillary rise of saline groundwater posed a potential threat to development over 30–40% of the city site, requiring either reconsideration of the established plans or the adoption of control measures (Cooke et al, 1982: 189). Figure 2.2 shows a map of the urban site and the aggressive ground conditions existing there.

In the case of Dubai, a survey was made of the Mina Jebel Ali port site, where a major new dock complex was planned (Figure 2.3). The depth to the

24

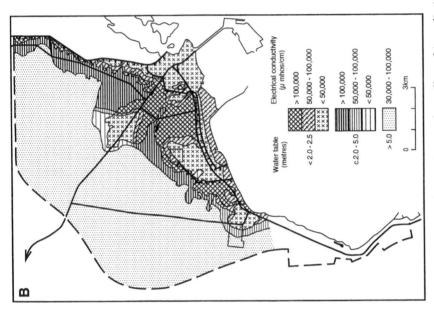

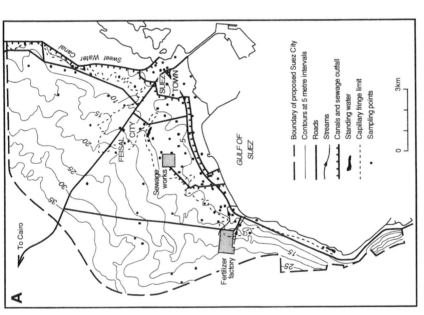

Figure 2.2. Suez City, Egypt. Aggressive ground conditions. (A) Groundwater sampling points and capillary fringe limit. (B) Groundwater hazard intensity based on water-table depth and electrical conductivity. Modified from Cooke et al (1982, figures v.17 and v.19) and reproduced by permission of the United Nations University Press. 1 μmhos cm⁻¹ = 0.1 mS m⁻¹.

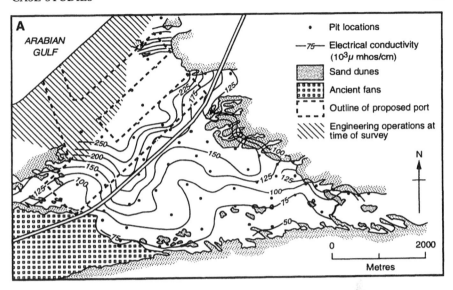

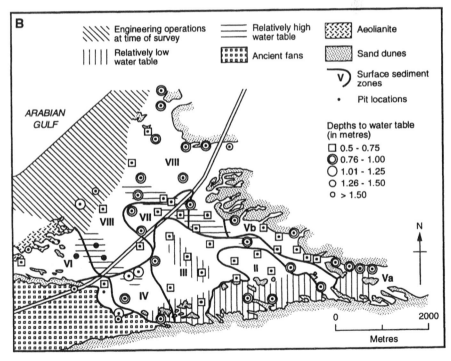

Figure 2.3. Salt weathering conditions in Dubai, United Arab Emirates. (A) Electrical conductivity pattern for groundwaters. (B) Height of water-table. Modified after Jones, 1980, figure 7, p. 67 and Cooke et al (1982, figure v.16A). $1\,\mu$mhos cm^{-1}=0.1 mS m^{-1}

Figure 2.4. Building in Ras Al Khaimah, United Arab Emirates located on the edge of a sabkha where saline groundwater is close to the surface. The concrete and iron reinforcements are suffering from extreme decay. It is unlikely that this building, which was probably completed in the 1970s, was ever used (photograph by A. S. Goudie)

water-table was often no more than 0.5–1.5 m, salinity levels reached 340 000 μmhos cm^{-1} (up to six times the salinity of the adjacent Arabian Gulf), and large areas were underlain by gypsum-rich sediments. It was appreciated that in such an environment the design of the dock quay walls was of great significance. As Jones (1980: 66) reported

> Impervious walls would result in the raising of water-table and salinity values over much of the sabkha, while pervious walls could cause the reverse, with the result that some of the sedimentary gypsum could go into solution and cause subsidence. In addition the survey revealed that industrial development on the sabkha was likely to prove costly as buildings would have to be placed on a layer of fill up to 3m thick and would require full protective measures. Thus the slopes of the higher ground to north and south would be preferable to the invitingly flat surface of the sabkha.

Salt problems have also been uncovered further north at Ras Al Khaimah, which is one of the United Arab Emirates and borders the Arabian Gulf. In the west it consists of a low angle plain mantled by a range of unconsolidated Quaternary deposits including fans, sabkhas, dunes and coastal sediments. In this low-lying coastal zone there is abundant evidence of the decay of buildings at or near ground level. The afflicted buildings are both ancient and modern (Figure 2.4). The former include some nineteenth century defensive towers, the Islamic site at Julfar and, most spectacularly of all, a series of abandoned

Figure 2.5. In Ras Al Khaimah, United Arab Emirates, capillary migration of saline groundwater in low-lying situations has caused salt attack on many buildings. In this example, the bricks have decayed more than the mortar cementing them (photograph by A. S. Goudie)

Figure 2.6. This wall in Ras Al Khaimah, which is probably no more than four years old, is located on the sabkha surface. Damage is already evident to the breeze-blocks and the plaster coating (photograph by A. S. Goudie)

Figure 2.7. These breeze-blocks in Ras Al Khaimah have suffered from salt attack before they have been used. They have been placed on the salty sabkha surface, which is dominated by halite, and have absorbed salty solutions which have then evaporated, leading to salt crystallisation (photograph by A. S. Goudie)

buildings in the village region of Diyah, east of Rams (Figure 2.5). The latter include a large number of houses or walls, mostly built since 1988, along the Julfar–Rams road (Figure 2.6). The damage is particularly marked in breeze-blocks (Figure 2.7) and many walls have holes at their bases and in some cases have collapsed as a result of the damage. The problem increases as walls get older, but serious damage can be caused in 4–15 years in the hazard zone. The salt involved is sodium chloride, and it is derived from upward migration from a high groundwater table. Some may be derived from surface flooding at times of high storm surges and some is derived from the irrigation of gardens within house enclosures or from leaking sewage/drainage/water supply systems. Using a Protimeter moisture meter it was found that on walls the effect of salt and moisture was discernible up to 1.6 m above ground level.

ISLAMIC MONUMENTS IN UZBEKISTAN

In 1992 Akiner et al wrote (p. 257)

> The Central Asian Republic of Uzbekistan is famous for some of the most beautiful and important examples of Islamic architecture to be found anywhere in the world. In particular, the historic towns of Khiva, Bukhara and Samarkand include many outstanding individual buildings and ensembles of buildings. Many of these monuments, which in addition to their enormous historic value are of fundamental importance to the local tourist industry, face a serious threat from salt attack. Unless urgent action is taken to control the problem, the already serious damage will get worse: increasingly expensive repairs will be essential and some buildings are at real risk of collapse.

The monuments for the most part post-date the Mongol invasions of the early thirteenth century. Many of the finest monuments date from the fourteenth and fifteenth centuries, though many of those in Khiva date from the nineteenth century. The spread of irrigation has increased salt damage associated with elevated groundwater levels, and extensive damage occurs in a zone around the bases of buildings extending as high as 2.55 m above ground level. The evidence of such damage includes the fracturing and powdering of bricks, the extrusion of bricks from walls and the associated deformation of wall lines, the break up of original brick patinas, the removal of mortar from between bricks, ceramic tile fall, and the blistering and flaking of alabaster.

PUEBLO BUILDINGS OF THE AMERICAN WEST

Many of the Pueblo buildings of the American West, in Arizona and neighbouring states, are composed of a material called adobe. They are in effect made of sun-dried mud bricks. Attention was drawn to their history and state of deterioration by Hayden (1945: 373)

> Severe, occasionally disastrous, erosion of the bases of adobe walls in the arid and semiarid regions of the Southwest is a phenomenon which has long been observed and commented upon. In historic times, erosion of this type has caused collapse of adobe buildings in the Rio Grande Valley of New Mexico; about Tucson, Arizona, it's serious, as it is in the Salt River Valley. Evidence of similar erosion has been noticed in excavations of prehistoric massive adobe or caliche walls of Hohokam and Salado structures of the thirteenth and fourteenth centuries A.D. in the Salt and Gila River valleys of Arizona. Repair to check erosion and prevent collapse of walls was the primary purpose of Cosmos Mindeleff's stay at Casa Grande in the 1890s. Fewkes noticed similar cutting of standing walls at the nearby Adamsville site.

Hayden noted that most of the erosion was achieved by undercutting between 15 and 20 cm above the ground surface and took on a different morphology from that erosion caused by rain-washing. Samples taken from the zone of erosion showed a soluble salt content of 87 590 ppm, whereas the upper wall contained only 2810 ppm. Some of the salt may be derived from a high groundwater level, and Hayden suggested (p. 377) that continued and intensified canal irrigation of the Salt River Valley [sic] by the Salado people raised the water table, increasing more and more the moisture content of the soil, bringing about a concentration of salts at the surface and a very rapidly increasing erosion of wall bases.

More recently Brown et al (1979) have discussed the factors that affect the durability of adobe structures in Arizona and cite calcium and magnesium sulphates as the salts involved.

THE SPHINX AND OTHER SITES IN EGYPT

The great Sphinx of Giza, one of the great architectural and archaeological gems of the world, has shown major deterioration since it was first photographed in 1850 and, by one estimate, the loss of stone is occurring at the rate of about 30 cm per century (Selwitz, 1990: 854). It consists of a limestone head and base that are composed of relatively hard and durable stone, but in between there is a limestone of much lower resistance. The limestone contains significant amounts of sodium chloride and calcium sulphate. These appear to cause extensive flaking and the salts may be intrinsic to the Eocene limestone itself (Livingston, 1989). Not all the damage to the Sphinx is necessarily caused by salt attack. Acid rain, natural solution of the limestone, wind abrasion and seismic attack may also be involved.

Major problems have been encountered elsewhere in Egypt, most notably around Luxor, where changes in hydrological conditions resulting from the Aswan High Dam have been implicated. The construction of the dam and the changes in irrigation practices associated with it have led to an increase in the height of the groundwater table, greater stability in the height of the water-table which makes it more difficult for periodic flushing of salts, and an increase in water salinity owing to the high levels of evaporation in Lake Nasser. The effects of these three tendencies on the monuments around Luxor have been described thus (Smith, 1986: 505–506)

> The groundwater provides a constant source of moisture that seeps into temple sites and travels by capillary action up into the stone walls and pillars. The main cause of stone deterioration of the monuments results from a combination of raised groundwater level and the capillary transport of salt-laden water. The most apparent damage is the result of crystallisation of the salt, which gradually breaks up stone surfaces and reliefs. The soil conditions around Thebes also encourage water-related damage to the monuments. The Nile floods deposited layers of fine-grained alluvial soil several m thick. These soils have a higher capillarity than the stone walls of monuments, so that salts are transported upwards and accumulate on the surface of the walls. These are hygroscopic salts that also contribute to the deterioration of the stone material.
>
> The higher water table makes it easier for water to migrate to the surface of the sandstone from which the monuments were built. The water eventually evaporates but the salt remains, blistering stone surfaces and crumbling the stone. Salt crystals have been likened to a sleeping devil; only when moisture is added do they begin to act.

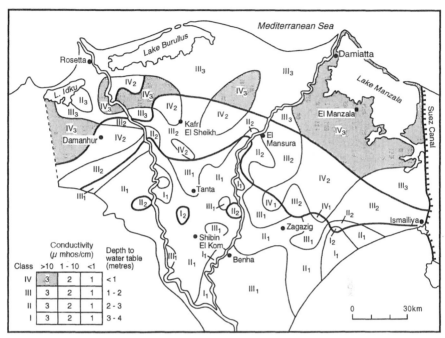

Figure 2.8. Predicted salt weathering hazard intensity in the Nile Delta of Egypt. Modified after Ibrahim and Doornkamp (1991, figure 5) and reproduced by permission of the authors. $1\,\mu$mhos cm^{-1}=0.1 mS m^{-1}

Similar problems were anticipated with respect to the Philae temples in the Aswan area (Voûte, 1962).

Salt attack on the walls of Cairo has long been appreciated, as described by Lucas (1915) in Hume (1925: 213)

> In the case of walls that have been plastered, the plaster is frequently forced bodily away from the wall, and in between the wall and the plaster a sheet of almost pure sodium chloride, sometimes one or even two millimetres in thickness, was found. In other cases small cavities in the mortar or in the stone were filled with a powdery mass of crystals of almost pure sodium chloride. In fact, sodium chloride was the chief constituent of all the efflorescences examined, though sometimes nitrates and sulphates were also present.

It is likely that there is also considerable potential for salt damage in the Nile Delta region. Groundwater is everywhere within 4 m of the surface and salt concentrations are high, especially in the northern portions bordering the Mediterranean Sea (Ibrahim and Doornkamp, 1991). Combining data on the depth to the groundwater and levels of salinity it is possible to map different classes of salt weathering hazard intensity within the Nile Delta (Figure 2.8). There are other parts of Egypt where salt weathering intensity may be considerable, including the gypsum- and halite-rich sabkhas that adjoin the Bitter-Lakes in the Suez Canal Zone (Wali, 1991)

SALT WEATHERING IN BAHRAIN

Bahrain is an island in the Arabian Gulf. Its northern end has been extensively developed, and there are large urban areas (e.g. Manama and Muharraq) and major industrial complexes. It was evident by the late 1960s and early 1970s that many highly expensive modern structures in Bahrain were undergoing rapid decay, and so the Bahrain Surface Materials Resources Survey was instituted to investigate the problem (Doornkamp et al, 1980). Large portions of the northern end of the island are low-lying and the groundwater is both shallow and saline. The climate is arid, temperatures are high, and the rates of evaporation (and capillary rise from shallow water-tables) are substantial. The survey thus concentrated on the hazard that groundwater posed to concrete buildings (Brunsden et al, 1979).

As a first step the extent of areas affected by capillary rise were mapped on the basis of three criteria — damp surface materials, 'puffy' ground and salt efflorescence. This facilitated the determination of the capillary fringe limit at the time of the survey (April 1975) (Figure 2.9). Secondly, estimates and observations were made of the height of capillary rise in various materials inland of the capillary fringe limit and from these data it was possible to predict that for all planning and engineering purposes the inland limit of the hazard could be taken as the 10 m contour (Figure 2.9c). Thirdly, an attempt was made to determine the spatial variability of the hazard within the hazard zone. The hypothesis was established that the aggressiveness of the ground is related directly and equally to the shallowness of the groundwater and the groundwater salinity. To resolve this hypothesis in spatial terms, data were collected at 147 sampling points for the depth of the water-table below the surface, the groundwater salinity as determined by electrical conductivity and the ionic concentrations in groundwater of Cl^-, SO_4^{2-}, Na^+, K^+, Ca^{2+}, and Mg^{2+}. The data were used to produce distribution maps.

The groundwater level map (Figure 2.9A) refers to the depth of the water-table below ground level (not below a horizontal datum) and the isolines are interpolated between sample points only. In general, the water-table depth increases inland, although at the capillary fringe limit and the 10 m contour its predicted depth varies considerably, mainly as a reflection of varying bedrock and superficial materials. The pattern is further complicated by geological inliers and dune fields. Estimates of the height of capillary rise indicate that this is normally less than 3.0 m above the water-table. Few areas below the 10 m contour are unaffected by capillary rise to near the surface. For its part, the conductivity map (Figure 2.9B) reveals not only a relatively low general electrical conductivity in the hazard zone (normally less than $3800 \mu mhos\, cm^{-1}$), but also five distinct areas of relatively high conductivity that are likely to be areas of potentially relatively high salt weathering hazard.

On the assumption that the intensity of salt weathering hazard is directly and equally proportional to both the shallowness of the groundwater and to

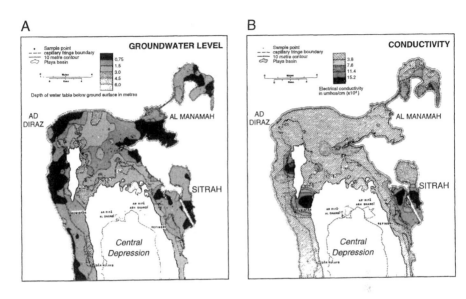

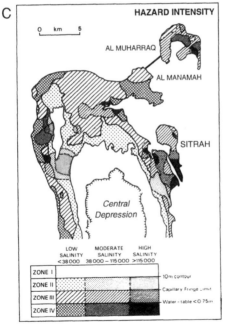

Figure 2.9. Salt weathering conditions in Bahrain. (A) Groundwater level. (B) Conductivity. (C) Hazard intensity. The salinity values are in μmhos cm^{-1}. Modified after Cooke et al (1982: 177, 178, 180) and reproduced by permission of the United Nations University Press

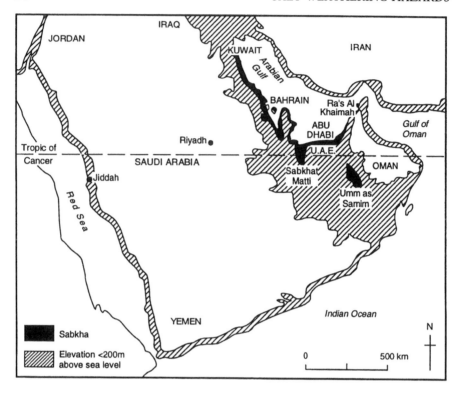

Figure 2.10. Sabkhas of the Arabian Peninsula and Arabian Gulf

the electrical conductivity of the groundwater, the distributions shown on Figure 2.9A and 2.9B were combined to produce a map that is a first approximation of hazard intensity (Figure 2.9C). This hazard map is a valuable basis for preliminary planning and engineering decisions, but it does have several obvious limitations: it is based on an, as yet, untested hypothesis; it is based on data for a single period of time (water-table levels are changing in both the long and short term, for instance); and the computer analysis involved a "smoothing" of information.

Sabkhas of the type found in Bahrain have been noted to be hazardous areas for salt attack in the United Arab Emirates and, given their considerable extent (Figure 2.10), it is likely that salt attack could be a problem along much of the western coast of the Arabian Gulf.

SALT DAMP IN ADELAIDE, AUSTRALIA

Salt attack on buildings has been noted from various locations in Australia, including the law quadrangle of the University of Melbourne, the 1860 police

Figure 2.11. Salt damp problems in Adelaide, South Australia as shown by efflorescence on the stonework of Glenelg parish church (photograph by H. A. Viles)

barracks at Bendigo and the Penal Settlement buildings (mainly 1835–1845) at Port Arthur, Tasmania. However, the situation appears to be particularly severe in Adelaide (Figure 2.11), where a Government of South Australia Salt Damp Research Committee indicated that more than 200 000 dwellings were affected and produced a map of the distribution of damage (see Blackburn and Hutton, 1980). Some of the buildings in the central area pre-dated the implementation of the statutory requirement for damp-proof courses (1881) and so lacked effective moisture barriers, but other problems result from poor drainage, faulty plumbing, careless watering of adjacent gardens and inadvertent bridging of damp-proof courses (e.g. when new paths, driveways, or flowerbeds are placed against a wall without due regard for the height and effectiveness of the damp-proof course). Some parts of the city also suffer from the effects of sea spray, but the problem is especially clear in low-lying areas where impermeable clay substrates have caused the development of a perched water-table lying within 3 m of the ground surface.

PIPELINE DECAY IN NAMIBIA

In the hyper-arid coastal desert of the Namib in Namibia the subsoils are rich in gypsum crusts and other evaporites (see, for example, Watson, 1983). To provide the tourist resort of Swakopmund and the large uranium mine at Rössing with a water supply, a large pipeline was constructed to carry water from the Omaruru delta aquifer.

Initially this pipeline was laid beneath the ground surface in a trench. Portions of the pipeline disintegrated just two years after installation (Bulley, 1986) because the pipes, consisting of concrete-coated steel casings reinforced with pre-stressed wire, were penetrated by salty moisture from the surrounding soil. Rapid corrosion and failure of the reinforcing wires took place, followed by disintegration of the pipe itself. Figure 2.12 shows the corroded nature of some of the pipes. The subsurface pipes have had to be replaced by coated steel pipes that are carried above the surface. The area also shows examples of the decay of concrete railway sleepers (Figure 2.13).

THE DISINTEGRATION OF ROADS AND RUNWAYS IN THE SAHARA

As Netterberg (1970: 3) pointed out, the presence of soluble salts has become recognised as a cause of certain types of bitumen road failure, particularly in dry regions

> The earliest symptoms of impending failure generally seem to occur during or soon after construction, and take the form of a slight powdering of the surface of newly constructed base before and/or after priming and/or white salt deposits in cracks in the surfacing and along the edges of the surfacing. Failure itself takes the form of blistering and cracking of thin surfacings, the lifting of localised patches of thicker surfacings with cracking of the uplifted patches, loss of density of the upper part of all bases, and powdering and loss of cohesion of the upper part of cement stabilised bases and some lime stabilised calcrete bases. In its mildest form the net-result is a weakening of the upper base, but in more severe cases a complete loss of bond between the base and the cracked surfacing results, with consequent "scabbing" of the surfacing and potholing.

He refers to problems that have been encountered in locations in western Australia, South Australia, South Africa, Namibia and northern India. A discussion of salt problems associated with calcrete roads in Australia is given by McNally (1995).

A detailed discussion of damage to roads and runways in the Algerian Sahara is provided by Horta (1985), who stresses that in this area the prime causes of damage are not hydrating salts (which have been implicated elsewhere), but halite (sodium chloride). He attributes salt damage to the heaving of the asphalt by salt whiskers growing vertically from the capillary pores of the road's base course. The heaving creates blisters which may reach heights of 10–15 cm depending on the pavement thickness. The problem is particularly serious in areas of high groundwater levels, as, for example, at wadi or sabkha crossings. He refers in particular to problems that have been encountered on National Road 6, in the Saoura Valley, and at the airport at Adrar.

The mechanisms by which such damage may occur have been summarised by Obika et al (1989) and are shown diagrammatically in Figure 2.14.

Figure 2.12. Water pipelines near Rössing, Namibia, which corroded and disintegrated after just two years in the salty soil of the Namib Desert (photographs by H. A. Viles)

Figure 2.13. These railway sleepers, made in 1959, were, 20 or so years later, severely damaged by salt attack. The location is the former track of the main railway line between Swakopmund and Walvis Bay in Namibia. The environment is extremely salty because of the extreme aridity of the climate and its proximity to the sea (photograph by A. S. Goudie)

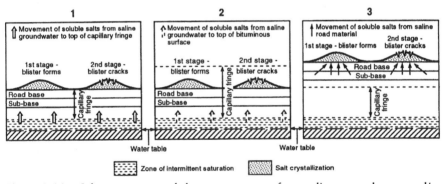

Figure 2.14. Salt movement and damage processes from saline groundwater and/or salt from construction material. (1) Pavement built within the capillary fringe. Salts can move in solution from the saline water-table to the underside of a relatively impermeable bituminous surface. Crystals may form in such a pattern to create forces sufficient to lift the surfacing. (2) Pavements built below a capillary fringe, but with no salt in the material. Salt can move in solution from the saline water-table to the surface of a relatively impermeable surfacing. Crystals form at the surface causing physical degradation of the bituminous surfacing. (3) Pavement built above a capillary fringe. Salts in the pavement material can move in solution even if above the capillary fringe. They crystallise at the underside of the relatively impermeable bituminous surface and, depending on their form, may lift the surfacing. From Obika et al (1989, figure 1)

SALT HEAVE AND CONCRETE CORROSION IN CALIFORNIA

The area between Long Beach and Newport Beach in the Los Angeles area of California, USA has many of the characteristics of a typical sabkha environment. It is low-lying, has high groundwater levels (between 2.9 and 3.2 m below the ground surface during drought years, but reaching to within centimetres of the surface in wet winters), high rates of evaporation and high solute concentrations in groundwater. Gypsum is widespread. The area originally consisted of swamps and salt marshes, but is now covered by concrete and asphalt. Building development has largely taken place since the 1960s, and during the mid-1980s the first buildings were showing clear signs of concrete corrosion and slab heaving. These problems have been recognised throughout the southern coastal plain of Los Angeles and Orange Counties (Robinson, 1995). During the middle and late 1980s millions of dollars were spent repairing damage caused by the corrosion and heaving of concrete house slabs, and modified construction techniques have had to be introduced. These include the use of sulphate-resistant cement, the use of stronger reinforcements in slabs, and the installation of membranes and metal weep screes to act as barriers to saline solutions.

THE NABATEAN CITY OF PETRA, JORDAN

In southern Jordan, to the east of the great Rift, lies the ancient city of Petra. It is notable for the large number of fine buildings that have been carved into the multi-coloured sandstones of the Ram Formation (Cambrian to Ordovician in age). The great bulk of the building and carving was undertaken by the Nabateans between around the sixth century BC and the first century AD. It is remarkable that in spite of the fact that many of the monuments are over 2000 years old they still in part display clearly and crisply many of the details of their working (e.g. chisel marks) and of their decoration. It is for this reason that Petra is Jordan's prime archaeological attraction and a magnet for tourists.

However, this splendid state of preservation is only partial, and the lower portions of many of the monuments show a substantial degree of decay (Figure 2.15). This is true of the lower portions of Al-Khasneh (The Treasury), parts of which have now been restored, Ed Deir (The Monastery), the Bab el-siq triclinium and obelisk tomb, the so-called royal tombs (The Tomb of Uneishu, the Urn Tomb, the Silk Tomb, the Corinthian Tomb and the Palace tomb), and various other tombs along the Wadi Farasa, including the Roman Soldiers' Tomb. There is also abundant evidence in the sandstone cliffs of Petra of the development of large numbers of cavernous weathering forms, ranging in size from small alveoles to huge tafoni. Similar features can be seen to be developing on some of the monuments.

One of the causes of the localised, but serious, weathering at the base of so many of the monuments is undoubtedly salt. Salt efflorescence is visible on

Figure 2.15. The Royal Tombs at Petra, Jordan show considerable amounts of weathering, particularly in their lower portions. The sandstones appear to be weathered by various salts, efflorescences of which are clearly evident in the foreground (photograph by A. S. Goudie)

most of the weathered areas and is accompanied by blisters and flaking. The area is dry, (mean annual rainfall about 190 mm), but winter season precipitation seeps out of the cliff faces into which the monuments are carved, then evaporates and precipitates salts. Albouy et al (1993) and Fitzner and Heinrichs (1991) have discussed the role of salt at Petra. More recently, Fitzner and Heinrichs (1994) have given details of the salts involved, drawing attention to halite, gypsum, potassium nitrate and magnesium sulphate.

Our own observations of the salinity levels associated with weathered and blistered outcrops at Petra indicate that the affected rock contains a large quantity of saline material. The conductivity of a $1:5$ mixture of powdered rock and deionised water from 14 locations (six from inside overhangs and eight from outside) ranged from 1610 to 46 300 μS cm^{-1}, with a mean of 20 158 μS cm^{-1}.

Figure 2.16. The cloisters of the Jeronimos Monastery, Lisbon have become badly blackened and weathered by the combined effects of air pollution, sea salt and lichens (photograph by H. A. Viles)

JERONIMOS MONASTERY, LISBON

Built in the sixteenth century, this fine and richly ornamented building, which was recognised as a world cultural heritage property by Unesco in 1982, illustrates the complexity of many urban salt weathering situations (Figure 2.16). The monastery, built in ornate style for King Manuel 1 in Lioz Stone (a local Cretaceous shelly limestone) is 200–280 m from the River Tagus in an area characterised by heavy road and rail traffic. As Lisbon is close to the sea, oceanic influences may also affect the building. In the winter of 1989–1990 part of the Chapterhouse roof collapsed, focusing attention on the decayed state of at least some of the building. Several styles of weathering are present on the monastery (Figure 2.17), including white, water-washed areas dominated by dissolution and surface retreat, black crusts resulting from the interaction of air pollution (especially sulphur dioxide and soot) with the stone, and lichen-encrusted areas. The black encrustations often peel off producing unsightly exfoliation and loss of architectural detail.

An extensive programme of work has been undertaken in recent years to describe the severity of decay, to explain its causes and relationship with microclimatic conditions and to suggest solutions. Vleugels et al (1992) have studied the geochemistry of runoff waters and crusts in particularly decayed parts of the building, which include the Southern Portal, San Vicente portal and cloisters. The crusts, up to 400 μm thick, are enriched with sulphur,

Figure 2.17. Detailed view of the upper part of the cloisters at the Jeronimos Monastery, Lisbon showing blackened, gypsum-encrusted areas, water-washed zones and the presence of higher plants

potassium, silicon, iron and magnesium. As shown in Table 2.2, ion chromatography reveals the dominance of sulphate (from air pollution), with much lower levels of chlorides (from maritime sources) and nitrates (presumed to be from fertilisers and air pollution). Phosphates and oxalates, present in relatively small amounts in some crust samples, are taken to indicate biological action, e.g. lichens, which can contribute to crust development. Sulphate makes up to 29% of the weight of the weathering crusts. Microclimatic studies help to explain these geochemical results, showing, for example, that in the winter rainy season the wind comes from the NW–SSW, bringing sea salt onto the monastery.

Table 2.2. Ion chromatography results from weathering crusts on Jeronimos monastery, Lisbon. After Vleugels et al (1992). Reproduced with permission of Elsevier Science

Crust sample	Cl^-	NO_3^-	SO_4^{2-}	PO_4^{3-}	Oxalate
White stone+ black crust	0.1	0.032	4.8	—	—
Brown–black crust	0.076	0.045	12.6	—	—
Brown–black crust	0.067	0.095	0.26	0.31	—
Brown crust	0.23	0.068	2.2	0.15	0.56
Black crust	0.19	0.22	29	0.074	0.089

All results expressed as concentrations in weight %.

SEA WALLS AT WESTON-SUPER-MARE, AVON, UK

The salty marine environment can pose a considerable hazard to sea walls and other engineering structures. As well as the erosive nature of storm waves which can attack sea walls periodically, causing major damage, salt weathering may create a more pervasive nuisance. Dated sea walls of varying exposure and orientation in turn provide a good natural laboratory for studying the variability of salt weathering intensity.

At Weston-Super-Mare a Carboniferous limestone sea wall, capped by sandstone coping stones, was completed in 1888. Lying 100 km upstream of the mouth of the Bristol Channel, Weston-Super-Mare experiences a tidal range of > 12 m and its westerly aspect means that it is exposed to winds with a transatlantic fetch. Major storm events breached the sea walls in 1903, 1981 and 1990. Many of the sandstone coping stones are now severely weathered into varying degrees of honeycombing. Mottershead (1994) has studied the weathering of the coping sandstones here, intensively mapping 862 stones in detail. The sandstone used is Forest of Dean stone, a Pennant sandstone of lower Carboniferous age. It is a massive, fine-grained, slightly calcareous sandstone, whose surface was rusticated by hammer dressing. Mottershead (1994) sampled four different sections of the sea wall (Figure 2.18) displaying different heights above sea level, degrees of exposure and orientations. A summary of his results is presented in Table 2.3. Exposure to salt spray and salt-laden winds from the south-west proved to be a more significant factor than the distance from the high water mark in explaining the degree of weathering (called weathering grade by Mottershead). The highest section of the walls (at Birkett) has a low mean weathering grade, suggesting that the vertical height is important as well as the horizontal distance to the high water mark. Chlorine contents of the top 5 mm of sandstone mirrored the trends in weathering grade, providing some support for the hypothesis that salt weathering has produced the honeycombing. Mottershead (1994) quotes nearby measurements of chlorine deposition on the island of Flatholm (about

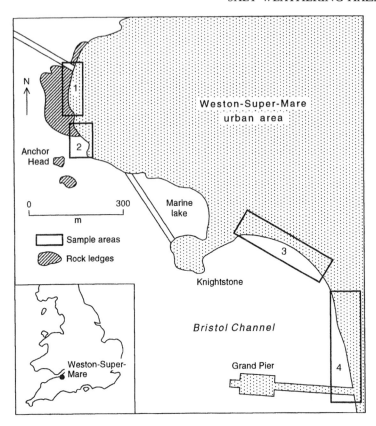

Figure 2.18. The location of Weston-Super-Mare, UK, showing positions of the sampling sites used by Mottershead (1994). Reproduced with permission

10 km offshore), which show about $9\,g\,m^{-2}\,yr^{-1}$ deposited, often in storm events.

Comparing the elevation of the weathered stone surfaces with the adjacent mortar (which tends to be more resistant to weathering), Mottershead (1994) derived a rate of about $1\,mm\,yr^{-1}$ of surface lowering averaged over the lifespan of the walls. Thus in just over 100 years honeycombing presumed to be caused by sea spray and splash, often of an episodic nature, has completely denuded some surfaces, removing over 10 cm depth of stone.

THE HISTORIC CENTRE OF OXFORD, UK

Oxford contains a wealth of historic buildings ranging widely in date, architectural style and function. Most of these historic buildings are faced with

LIBRARY, UNIVERSITY OF CHESTER

Table 2.3. Weathering and chloride contents of four sections of sea wall at Weston-Super-Mare, Avon, UK. After Mottershead (1994) and reproduced with permission

	Sites (see Figure 2.18 for locations)			
	Birkett (1)	Knightstone (3)	Anchor (2)	Royal Parade (4)
Elevation (m above MHWST)	8.5–9.5	2.5	2.8–3.5	2.5
Exposure	Exposed	Sheltered (W); exposed (E)	Exposed	Partly exposed
Foreshore at HWM	Rocky	None	Rocky	Sandy beach
Distance to HWM (m)	0–5	0	0–5	10–50
Orientation	W	SW	S–W	W
Weathering grade (mean values)*	1.95 ($n=25$)	Exposed (E) 6.12 ($n=31$); sheltered (W) 1.45, ($n=19$)	Exposed 6.45 ($n=9$); sheltered 2.55 ($n=8$)	4.5 ($n=55$)
Mean chloride content (mg l^{-1})	16.9 ($n=9$)	13.6 (in W); 20.2 (in E); 16.9 (overall) ($n=12$)	29.7 ($n=8$)	16.9 ($n=12$)

*Weathering grade scale: 0=no visible weathering forms; 1=isolated circular pits; 2=pitting >50% of area; 3=honeycombing present; 4=honeycombing >50% of area; 5=honeycombing with some wall breakdown; 6=honeycombing partially stripped; 7=honeycombing stripping >50% of area; 8=only reduced walls remain; and 9=surface completely stripped.

limestone. From the medieval period to around the eighteenth century, local limestones of a rich honey colour from Wheatley, Headington and further away in the Cotswolds were the dominant stones used. Headington stone came in two varieties: Headington Hard, a durable stone used for plinths; and Headington freestone, a much softer stone used for ashlar and decorations. Headington freestone became notorious for its lack of resistance to weathering in polluted atmospheres (Figure 2.19) and by the nineteenth century stone was commonly brought in from other areas of Britain (notably from Bath and Clipsham), as well as from France, to repair damaged stonework and for use in new constructions. Despite being a university town with little heavy industry, nineteenth century and early twentieth century Oxford air was polluted with coal smoke and sulphur dioxide. Photographs from the turn of the century show heavily blackened, gypsum- and soot-encrusted Headington freestone walls with characteristic unsightly blistering as the crust peels away from the underlying damaged stone layer.

Extensive repair and refacing of buildings has taken place in Oxford from the late nineteenth century onwards, and by the 1970s many of the historic buildings were clean and unweathered. Despite the general improvements in air quality in English cities by this time, and the use of generally better quality, more durable stone, stone decay has continued and salts from a range of sources are still implicated. Traffic is now a major nuisance within the historic city centre, and increased emissions of nitrogen oxides and other pollutants seem to be causing localised blackening and decay of stone surfaces near busy

Figure 2.19. The Bodleian Library, Oxford was built out of local Headington limestone which has become badly blackened and encrusted through sulphation (photograph by H. A. Viles)

roads. Road salt applications in icy winters have also apparently lead to problems on the lower sections of some roadside walls. Much old, encrusted and blistering stone still remains and its vulnerable state can easily be worsened by renewed weathering by salts. A good example is provided by the wall around Worcester College, which faces a busy street and is currently showing catastrophic decay.

CONCLUSIONS

From the range of case studies presented on the previous pages it is clear that salts pose a major hazard to new and old buildings and engineering structures (including roads and sea walls) in many locations. Urban, arid and coastal environments are particularly at risk. It is not simply natural stone that is affected, although salts can attack a wide range of stones including some, such as granite, which are often durable in the face of other environmental factors. Brick, concrete, adobe tiles, asphalt, ferrous metals and other porous or reactive materials are also prone to attack.

From case studies presented here, and many other studies such as those listed in Table 2.1, it can be seen that a wide range of salts can be responsible for devastating weathering, and that these salts commonly come from one or more of the following sources: airborne (including sea spray), groundwater, irrigation water, the building materials themselves and mortars. Furthermore, these case studies indicate that in many cases salt weathering (often in more than one form, utilising different salt sources) works in association with a range of other weathering processes, and untangling the contributions of each process can be difficult. Several of the case studies also illustrate the long-term effects of salt weathering, and how once salts become emplaced within building materials they can remain active, or become reactivated many years later.

Finally, these case studies illustrate the real costs of the salt hazard: the fabric of affected historic monuments needs repairing and restoring; structural damage to engineering structures may occur; and many new buildings and structures built without proper appreciation of the salt hazard have very short lifespans. Methods to counteract salt weathering (examined in more detail in Chapter 7) range from preventative through protective to restorative, and all have their associated costs. Salt weathering does not only affect rich, Western nations; it operates in many poorer countries as well, affecting their income-generating cultural heritage and basic essential infrastructure. Salt weathering therefore poses a global challenge.

3 Nature of Salts Involved in Salt Weathering and Sources of Moisture

INTRODUCTION

In this chapter we consider the variety of salts implicated in salt weathering and how such variety can be accounted for. We show that salts are indeed diverse in type and that there are many controls which determine which salts occur in which locations (see Warren, 1989, and Melvin, 1991, for reviews). We also investigate the different sources of moisture which aid salt weathering. Consideration of both the diversity of salts and moisture sources allows us to suggest under which environmental conditions salts are particularly effective agents of weathering.

Salts are ionic compounds formed between cations (except H^+) and the anions of acids. As all chemistry students know, the reaction between an acid and a base produces water and a salt. Common types of salt include chlorides, sulphates, nitrates and carbonates. Salts possess many qualities which render them important agents of weathering. Salts are soluble and can dissolve and recrystallise; many are hygroscopic or able to take up water from the air, and in some cases this produces hydrated forms with the possibility of changes of state between hydrous and anhydrous forms. In extreme cases hygroscopicity of salts can result in deliquescence, or the ability of the salts to dissolve in the water they have taken up. Salts also participate in other chemical reactions in the presence of water, reacting with minerals and rock surfaces. Finally, salts may be produced biologically by a range of organisms through biochemical processes.

In general terms we can describe the various cycles involved in salt accumulation in terms of five main situations that have been adopted by the FAO and Unesco (Szabolcs, 1979: 32):

1. Continental cycles connected with the redistribution, movement and accumulation of carbonates, sulphates and chlorides derived from the weathering of different types of rocks in inland regions that have inward flowing drainage.
2. Marine cycles connected with the accumulation of marine salts, mainly sodium chlorides, on the coastal plains of dry lowlands and along the shores of shallow bays.

3. Delta cycles. These are widespread and of great importance to humankind. River delta areas have been extensively used for irrigation since ancient times (the Amu Darya, the Tigris–Euphrates, the Hwang-ho, the Nile). Their cycles are characterised by a complex combination of movement processes and by the accumulation of salts carried either from the continent by rivers and delta-valley ground systems, or sometimes from the sea.

4. Artesian cycles connected with the evaporation of deep underground waters forced towards the surface through tectonic fractures and other routes.

5. Anthropogenic cycles resulting from errors in the economic activities of humans or from ignorance of the laws of salt accumulation (e.g. salinisation of irrigated soils through a rise in the water-table, salinisation of meadows through over-pasturing, irrigating with mineralised waters and flooding of waters from mining shafts, petroleum waters or other industrial waste waters).

Such cycles, along with the ever-increasing influence of air pollution which, as we noted in Chapter 1, provides an important source of salts or chemical precursors of salts, lead to a range of different types of accumulation of salts within a range of environmental settings. Efflorescence is a term generally used to describe the process whereby salts come to a surface (be it a sedimentary body, rock surface or building structure) and crystallise on it. Hydrated salts may lose water on exposure to air, and thus form a fine, powdery deposit on the surface. In some cases, hydrated salts can effloresce within a porous body, and such a phenomenon is often called subflorescence (Amoroso and Fassina, 1983: 29–33), especially in the context of porous building materials. Efflorescences may reach a considerable thickness, forming a clear crust which may have a significantly different colour, hardness and chemical reactivity to the underlying surface. Over a geological time-scale such crusts may produce distinct evaporite beds. The ability of salts to effloresce has been commercially harnessed by human societies over the millennia to harvest salt and purify water within salt pans and desalination plants.

NATURE OF SALTS

Nature of efflorescences in polar regions

The presence of efflorescences on naturally occurring rock surfaces has often been found to be a feature of outcrops undergoing salt weathering, and so the analysis of efflorescences gives an indication of the salts that may be involved in the process. An extensive literature survey indicates that efflorescences have a wide variety of mineralogies (Table 3.1) and that by implication a large number of different salt types may be involved in salt weathering. The information provided here builds on an earlier survey (Goudie and Cooke, 1984).

Table 3.1. Major evaporite minerals and their chemical composition

Name of mineral	Chemical formula
Alunite	$(K,Na)Al_3(SO_4)_2(OH)_6$
Ammonia niter	NH_4NO_3
Anhydrite	$CaSO_4$
Antarcticite	$CaCl_2 \cdot 6H_2O$
Aphthitalite (glaserite)	$K_2SO_4 \cdot (Na,K)SO_4$
Aragonite	$CaCO_3$
Arcanite	K_2SO_4
Bassanite	$CaSO_4 \cdot \frac{1}{2}H_2O$
Bischofite	$MgCl \cdot 6H_2O$
Bloedite (astrankanite)	$Na_2Mg(SO_4)_2 \cdot 4H_2O$
Burkeite	$Na_2CO_3 \cdot 2Na_2SO_4$
Calcite	$CaCO_3$
Carnallite	$KMgCl_3 \cdot 6H_2O$
Darapskite	$Na_3(NO_3)(SO_4) \cdot H_2O$
Dolomite	$CaMg(CO_3)_2$
Epsomite	$MgSO_4 \cdot 7H_2O$
Ettringite	$Ca_6Al_2(SO_4)_3(OH)_{12} \cdot 26H_2O$
Gaylussite	$CaCO_3 \cdot Na_2CO_3 \cdot 5H_2O$
Glauberite	$Na_2Ca(SO_4)_2$
Gypsum	$CaSO_4 \cdot 2H_2O$
Halite	$NaCl$
Hanksite	$Na_2K(SO_4)_9(CO_3)_2Cl$
Heptahydrite	$Na_2CO_3 \cdot 7H_2O$
Hexahydrite	$MgSO_4 \cdot 6H_2O$
Humberstonite	$K_3Na_7Mg_2(SO_4)_6(NO_3)_2 \cdot 6H_2O$
Huntite	$Mg_3Ca(CO_3)_4$
Hydrohalite	$NaCl \cdot 2H_2O$
Jarosite	$KFe_3(SO_4)_2(OH)_6$
Kainite	$MgSO_4 \cdot KCl \cdot 3H_2O$
Kalicinite	$KHCO_3$
Kieserite	$MgSO_4 \cdot H_2O$
Langbeinite	$K_2Mg_2(SO_4)_3$
Leonhardtite (Starkeyite)	$MgSO_4 \cdot 4H_2O$
Leonite	$MgSO_4 \cdot K_2SO_4 \cdot 4H_2O$
Loewite	$Na_{12}Mg_7(SO_4)_{13} \cdot 15H_2O$
Magnesite	$MgCO_3$
Magnesium calcite	$(Ca,Mg)CO_3$
Mirabilite	$Na_2SO_4 \cdot 10H_2O$
Nahcolite	$NaHCO_3$
Natron (natrite)	$Na_2CO_3 \cdot 10H_2O$
Niter (nitrokalite)	KNO_3
Nitratine (soda nitre)	$NaNO_3$
Nitrocalcite	$Ca(NO_3)_2 \cdot 4H_2O$
Nitromagnesite	$Mg(NO_3)_2 \cdot 6H_2O$
Northupite	$Na_3Mg(CO_3)_2Cl$

Continued

Table 3.1. *continued*

Name of mineral	Chemical formula
Pentahydrite	$MgSO_4 \cdot 5H_2O$
Picromerite	$K_2Mg(SO_4)_2 \cdot 6H_2O$
Pirssonite	$CaCO_3 \cdot Na_2CO_3 \cdot 2H_2O$
Polyhalite	$2CaSO_4 \cdot MgSO_4 \cdot K_2SO_4 \cdot 2H_2O$
Rinneite	NaK_3FeCl_6
Rosenite	$FeSO_4 \cdot 4H_2O$
Sanderite	$MgSO_4 \cdot 2H_2O$
Shoenite (picromerite)	$MgSO_4 \cdot K_2SO_4 \cdot 6H_2O$
Shortite	$2CaCO_3 \cdot Na_2CO_3$
Sylvite	KCl
Syngenite	$CaSO_4 \cdot K_2SO_4 \cdot H_2O$
Tachyhydrite	$CaCl_2 \cdot 2MgCl_2 \cdot 12H_2O$
Thenardite	Na_2SO_4
Thermonatrite	$Na_2CO_3 \cdot H_2O$
Trona	$Na_3(CO_3)(HCO_3) \cdot 2H_2O$
Ulexite	$NaCaB_5O_9 \cdot 8H_2O$
Vanthoffite	$MgSO_4 \cdot 3Na_2SO_4$
Weddellite	$CaC_2O_4 \cdot 2H_2O$
Whewellite	$CaC_2O_4 \cdot H_2O$

The first area we consider is the dry valley terrain of Antarctica, where salt weathering has been claimed to be "the dominant erosive process" (Wellman and Wilson, 1965: 1097). Its ice-free areas, characterised by extreme cold and aridity, receive some precipitation in the form of snow, but much is lost from the ground surface by sublimation. Parts of the dry valleys of the McMurdo Sound area are so dry that, almost uniquely (though see Levy, 1977) on account of its deliquescence, a hydrated form of calcium chloride ($CaCl_2 \cdot 6H_2O$ — antarcticite) is known from Don Juan pond (Harris et al, 1979). Some of the other lakes of the dry valleys contain other salts (Ball and Nichols, 1960). One of the most striking features of the ice-free areas is the presence, on a wide range of rock types, of efflorescences (Wand, 1995). Among the important types (see Table 3.2) are: gypsum, halite, thenardite/mirabilite, epsomite, hexahydrite, bloedite, soda nitre and sylvite. Such efflorescences have been observed for many decades (Debenham, 1920) and many workers have noted the breakdown of rocks where efflorescences are present (Prebble, 1967; Selby and Wilson, 1971). For example, at McMurdo Strait, Cailleux (1968) found that mirabilite was the effective salt, whereas near the Lower Rennick glacier Dow and Neall (1974) attributed the weathering to calcite, gypsum and magnesium carbonate, and in the Taylor Dry Valley Johnston (1973) invoked halite.

Wilson (1979) has tried to explain the spatial variation in salt types by reference to latitudinal zonation brought about by differences in the degree of

Table 3.2. Efflorescences in Antarctica

Location	Reference	Nature of deposit
Shackleton Glacier	Claridge and Campbell (1968)	Gypsum, thenardite, epsomite, hexahydrite, bloedite, darapskite, soda nitre
Mount Erebus and Ross Island	Jones et al (1983)	Halite, mirabilite, thenardite, gypsum, alunite, calcite
Rennick Glacier	Dow and Neall (1974)	Calcite, gypsum, magnesium carbonate
Victoria Valley	Johannesson and Gibson (1962)	Calcium sulphate, calcium carbonate, sodium sulphate, sodium nitrate, sodium chloride, sodium iodate
Marble Point (McMurdo Sound)	Ball and Nichols (1960)	Halite and thenardite
Dunlop Island	Nichols (1963)	Halite and thenardite
Molodezhnava	McNamara and Usselman (1972)	Calcite and halite
Victoria Valley	Gibson (1962)	Soda nitre, epsomite, gypsum, mirabilite, sodium iodate
Taylor Dry Valley	Johnston (1973)	Sodium chloride
Vestfold Hills	McLeod (1964)	Mirabilite and halite
Ohio Range	Dort and Dort (1970a, 1970b)	Hexahydrate and gypsum
Hobbs Glacier	Dort and Dort (1970a, 1970b)	Mirabilite
Sør Rondane Mts	Dort and Dort (1970a, 1970b)	Gypsum
Syoura	Dort and Dort (1970a, 1970b)	Mirabilite, thenardite, gypsum, halite, epsomite, bloedite and sylvite
Darwin Glacier (McMurdo Sound)	Dort and Dort (1970a, 1970b)	Gypsum, thenardite, halite, calcite and epsomite
Cape Dennison (George V coast)	Dort and Dort (1970a, 1970b)	Mirabilite
Black Island (Ross Ice Shelf)	Dort and Dort (1970a, 1970b)	Mirabilite
McMurdo Region	Keys and Williams (1981)	Thenardite, gypsum, halite, calcite, darapskite, soda nitre, mirabilite, bloedite, epsomite and hexahydrite
Victoria Land (Darwin Mts)	Miotke and Hodenberg (1980)	Calcite, mirabilite, thenardite, gypsum and halite

Continued

Table 3.2. *continued*

Location	Reference	Nature of deposit
Sør Rondane Mts	Matsuoka (1995)	Gypsum, thenardite, epsomite, jarosite, bloedite, nitratine, calcite, dolomite and halite
Schirmacher Oasis (Queen Maud Land)	Wand (1995)	Calcite, aragonite, burkeite, epsomite, gypsum, halite, hexahydrite, huntite, jarosite, malachite, pickeringite, rosenite, slavikite, trona

deliquescence of different salts, with sodium sulphate tending to occur at the highest altitudes, sodium chloride in intermediate positions, and calcium chloride and magnesium chloride in the lowest closed depressions.

Distance from the sea is also important as wind-blown marine salts are important in many parts of Antarctica (Wand, 1995), although halite appears to be absent from the Sør Rondane mountains (Matsuoka et al, 1996), probably because katabatic winds prevent the landward intrusion of sea salt. Just as there are clear examples of salt weathering associated with efflorescences in Antarctica (Miotke and Hodenberg, 1980), so in the drier areas of the Arctic a similar association has been proved (Table 3.3). On Ellesmere Island, for example, Watts (1981) related rock disintegration and

Table 3.3. Efflorescences in the Arctic

Location	Reference	Nature of deposit
Mesters Vig, NE Greenland	Washburn (1969)	Calcite, aragonite and trona
Prince Patrick Island, Queen Elizabeth Island	Tedrow (1977)	Thenardite
Ellesmere Island	Watts (1981)	Thenardite, calcium carbonate and sodium chloride
Søndre Strømfjord, Greenland	Hansen (1970)	Sodium sulphate and magnesium sulphate
Inglefield Land, Greenland	Tedrow (1970), Nichols (1969)	Chlorides of magnesium, calcium, sodium and potassium
Thule area, Greenland	Nichols (1969)	Gypsum
Peel River, Yukon	Traill (1970)	Epsomite
Ellesmere Island (Hazen Lake area)	Traill (1970)	Thenardite

tor development to the presence of efflorescences composed of thenardite, calcium carbonate and sodium chloride. The Arctic sites show a similarly wide range of efflorescence types as Antarctica.

Efflorescences in other parts of the world

Outside the polar regions efflorescences occur widely, notably in arid and semi-arid areas, and especially on shale outcrops. Sulphate minerals (gypsum, epsomite, bloedite and hexahydrite) form as oxidised surface water interacts with sulphides in the shale (Smoot and Lowenstein, 1991: 277). Their association with rock weathering has again often been reported (see, for example, Höllermann, 1975; Goudie, 1984). A notable feature of Table 3.4 is the relatively common occurrence of sodium and magnesium sulphates in their different forms (mirabilite, thenardite, epsomite and hexahydrite). In North Dakota, USA, the presence of magnesium sulphate efflorescences appears to be particularly evident in areas of dolomitic glacial drift (Keller et al, 1986). In the Negev epsomite efflorescences, which are widespread, seem to be derived from waters issuing from Cretaceous dolomites and gypsiferous marls (Mazor and Mantel, 1966).

One very special type of efflorescence is the sodium nitrate material ("caliche") which impregnates large areas of the Atacama Desert: it occurs in all topographic situations from the tops of hills and ridges to the centres of the broad valleys, and impregnates fractures and replaces all rock and sediment types (Eriksen, 1981; Searl and Rankin, 1993). The Atacama nitrates represent an unusual spatial segregation of salts due chiefly to the capillary migration of residual solutions away from a saturated location to an extent determined by the relative mobility of anions in solution (e.g. Mueller, 1968).

Salt lake chemistry and mineralogy

We now move on to consider the variable chemistry and mineralogy of lake evaporites from different parts of the world. We start with Australia, where many analyses of salt lake chemistry are available, and where the weathering of rocks by salt in lacustrine (Jutson, 1918) and coastal (Dunn, 1915) environments has been known to take place for a long time (Figure 3.1). The majority of the lakes, both in mainland Australia and in Tasmania, are dominated by sodium and chloride ions. This applies to the crater lake at Keilambete (Bowler, 1970) as much as to the non-volcanic lakes of Tasmania, Victoria, South Australia and Western Australia. There are exceptions to this general pattern, with thenardite being known from a group of isolated salt lakes between Ayers Rock and Erldunda and to the east of Lake Mackay in Northern Territory (Wells, 1976). In general, however, the uniformity in salt lake water chemistry over such enormous areas is remarkable, and possible explanations in terms of the role of cyclic salts and the redistribution of connate salts are returned to later in this chapter.

Table 3.4. Efflorescences from non-polar areas

Location	Reference	Rock type	Minerals
Fort St John, British Columbia	Traill (1970)	Shale	Epsomite and mirabilite
South Ontario	Traill (1970)	Shale and dolomite	Epsomite
Crow Lake, Ontario	Traill (1970)	Serpentine	Epsomite
Montreal, Canada	Traill (1970)	Utica shale	Epsomite
Upper Colorado Basin, USA	Whittig et al (1982)	Mancos shale	Epsomite, hexahydrite, pentahydrite, mirabilite
Arjanjuez, Spain	Gumuzzio et al (1982)	Alluvium	Sodium and magnesium sulphate
Konya Basin, Turkey	Driessen and Schoorl (1973)	Alluvium	Mirabilite, bloedite, thenardite, epsomite and halite
Karakoram Mts, Pakistan	Goudie (1984)	Granite, gneiss, schist, marble, limestone, shale, sandstone and volcanic rocks	Hexahydrite and gypsum
Diablo and Templor range of California, USA	Murata (1977)	Shale	Bloedite and hexahydrite
Capitol Reef, Utah, USA	Mustoe (1983)	Triassic and Jurassic sandstones	Gypsum, hexahydrite, bloedite and natron
Ohio, USA	Foster and Hoover (1963)	Dolomite	Hexahydrite
Tucuman, Argentina	Porto (1977)	—	Mirabilite and other sulphates
Tenerife	Höllerman (1975)	Volcanic rocks	Sodium chloride, sodium sulphate, calcium sulphate, (+$MgCl_2$, KCl, + $CaCO_3$)
Sind, Pakistan	Goudie (1977)	Alluvium and bricks	Thenardite, aphthitalite, burkeite, Na_2CO_3
Moradabad, India	Auden et al (1942)	Alluvium	Sodium carbonate and sodium sulphate
Meerut, India	Auden et al (1942)	Alluvium	Sodium carbonate and sodium sulphate
Khairpur District, Sind, Pakistan	Cotter (1923)	Alluvium and sand	Na_2CO_3, $NaHCO_3$, + some NaCl + Na_2SO_4

Continued

Table 3.4. *continued*

Location	Reference	Rock type	Minerals
Atacama Desert, Chile	Eriksen (1981)	All rock types	$NaNO_3$ + halite, glauberite, bloedite, gypsum and anhydrite, darapskite and humberstonite
Basin and Range Province, USA	Kirchner (1996)	All rock types	Halite, soda nitre, thenardite, gypsum, nachcolite, bassanite, thermonatrite
Southern Niger	Ducloux et al (1994)	Alluvium and sand	Konyaite, hexahydrite, epsomite, gypsum
Alberta, Canada	Kohut and Dudas (1993)	Drift	Sodium sulphate, konyaite and bloedite
Jaquetepeque, Peru	Mees and Stoops (1991)	Alluvium	Burkeite, eugsterite, halite, northupite, pirrsonite, thenardite and trona
Negev Desert, Israel	Mazor and Mantel (1966)	Calcareous and argillaceous rocks	Epsomite and halite
Asama, Japan	Matsukura and Kanai (1988)	Pumice flows	Halite, gypsum, hexahydrite, mirabilite

Salt lakes are also an important component of the landscape of arid America, especially in the Basin and Range Province. The chemistry of the lake basins shows a wide range of dominant ions. Lakes dominated by sodium carbonate and bicarbonate include Mono (California), an important group in Oregon (Abert, Harvey, Summer Lake, Alkali Valley) and Washington (Lenore and Soap), and some in British Columbia (e.g. Goodenough). Playa lakes dominated by sodium sulphate occur in the Laramie area of Wyoming (Kolm, 1982), in North Dakota and in Saskatchewan (Slezak and Last, 1985). Lakes dominated by magnesium sulphate occur in British Columbia and Saskatchewan, while many of the basins of southern California are dominated by sodium chloride.

However, within individual basins there is clear evidence of a zonation of salts. For example, in the Bonneville Salt Flats of Utah (Lines, 1979) three main zones have been recognised: an outer carbonate zone of authigenic clay-sized carbonate minerals; an intermediate sulphate zone composed mainly of gypsum; and an inner chloride zone composed dominantly of halite. Around Deep Springs Lake (California) the sequence of key minerals from the playa margin to the centre is reported by Jones (1965) as being calcite and/or aragonite, dolomite, gaylussite, thenardite and burkeite. In Saline Valley

Figure 3.1. Rock outcropping on the shoreline of a salty pan in Western Australia. Salt attacks such rock outcrops and in due course may produce a flat surface of haloplanation (photograph by A. S. Goudie)

(California) the sequence from periphery to centre is gypsum, gypsum and glauberite, glauberite, and glauberite and halite (Hardie, 1968). In Death Valley (Hunt, 1975) there is also an orderly horizontal arrangement of salt zones, with a chloride zone in the centre, a sulphate zone in the intermediate areas (with minerals such as trona, thermonatrite, thenardite and gypsum) and the carbonate zone at the outer edge of the salt pan, composed mainly of silt and sand, but with tiny crystals of calcite. Severe salt weathering has been noted in the sulphate zone near Badwater (Goudie and Day, 1980). The orderly arrangement of salt zones reflects, in most instances, the relative solubilities of salts and the sequence in which they crystallise from solution.

The lakes also show a variability in salt mineral zonation vertically, reflecting both changes in the environmental and sedimentary history of the basin and differences in the solubility of different salts. For example, at Deep Springs Lake the sequence upwards from basal carbonate muds is nahcolite, thenardite, burkeite, trona, halite and sylvite.

Most of the salt lakes of South America occur in basins within the great western cordillera. In Chile a comprehensive survey by Stoertz and Ericksen (1974) has revealed that sodium carbonate (trona) is rare in the Chilean salars, and that there are two main types of crust in most basins: an inner, hard evaporative crust zone composed dominantly of halite and a soft outer zone consisting chiefly of gypsum, mirabilite (or thenardite) and ulexite.

In the Bolivian Andes the basins show rather more variety (Risacher, 1978). Although in five of the seven lakes for which data are given the principal mineral deposits are said to be halite, gypsum, mirabilite (thenardite) and ulexite, there are two basins (Cachi Laguna and Collpa Laguna) where the dominant minerals are natron and thermonatrite.

In central Mexico there are two main types of water present. Volcanic crater lakes (La Piscina de Yunria and Hoya Rincon de Parageneo) are rich in sodium and carbonate, while the other basins are dominated by sodium chloride.

The most important lakes in Africa are those of the Rift Valley in the east of the continent. Many of these include sodium carbonate and bicarbonate lakes (e.g. Natron, Turkana, Bogoria, Nakuru, Magadi, Chitu and Shala), though Kivu is dominated by magnesium carbonate and bicarbonate, and the Katwe Volcanic Lakes in Uganda are dominated by sodium chloride (Arad and Morton, 1969).

Trona is also dominant in some of the small playa lakes of the Kanem area of the Chad Basin (Eugster and Maglione, 1979), though there are some areas with sodium sulphate and halite. Trona also occurs in some of the small basins of the Libyan sand sea in Libya (Goudharzi, 1970). Elsewhere in Libya, notably near Edri and Marada, halite is dominant, and this also applies to the Wadi Natrun lakes to the west of the Nile Delta in Egypt (Larsen, 1980). Severe salt weathering of volcanic rocks on the shores of African carbonate lakes has been noted, including the Galla Lakes of southern Ethiopia, Lake Chew Bahir (Ethiopia and Kenya) (Figure 3.2), Lake Nakuru and Lake Magadi (Figure 3.3).

One of the largest salt lakes in southern Africa is the Etosha Pan in Namibia, where the main constituent is halite, but other salts are thenardite, burkeite and probably also trona and hanksite (Buch and Rose, 1996).

Most of the salt lakes of Asia appear to be dominated by sodium chloride, including the Aral, Caspian and Dead seas, and most of the lakes of Anatolia, the Crimea and the Tibetan Plateau. However, as with many of the large American lakes, studies have shown that even in sodium chloride dominated lakes there is a clear zonation of precipitation of other salts. For example, the Karabogaz Gulf, adjacent to the Caspian Sea, has zones of halite, gypsum, glauberite, mirabilite and epsomite deposits (Strakhov, 1970: 171). Details of salt lakes in China and Mongolia are provided by Zheng et al (1993) and Egorov (1993).

In the ice-free regions of Antarctica there are frequent occurrences of saline lakes in enclosed basins. Their chemistry is complex and variable (Campbell and Claridge, 1987). Some contain calcium chloride in the form of antarcticite ($CaCl_2 \cdot 6H_2O$).

Coastal lagoons and sabkhas

In many coastal regions lagoons may form where a body of marine water becomes more or less permanently separated from the sea by a barrier such as

A

B

C

Figure 3.2. Lake Stefanie (Chew Bahir) in southern Ethiopia is an ephemeral and sporadically salty rift valley lake. The volanic rocks on its shore show clear signs of progressive haloplanation on moving from high (A) through medium (B), to low (C) levels. The aggressive salt is probably predominantly trona (sodium carbonate) (photographs by A. S. Goudie)

a sand spit. Where evaporation or other concentrative processes occur such lagoons become highly saline. Similarly, along arid coastlines of low relief, coastal supratidal sabkha plains may develop, forming a distinctive coastal wetland type in which evaporation produces high salt concentrations in surface and ground waters and sediments. For example, along the northern, Mediterranean coast of the Sinai peninsula, Egypt, the Bardawil lagoon occupies around 600 km^2, reaching a maximum of 3 m in depth. The lagoon is separated from the Mediterranean by a sand barrier and has salinities up to three times those of normal seawater (Levy, 1980). Most of the lagoon brines

Figure 3.3. The volcanic lavas on the margins of the highly saline Lake Magadi in Kenya have been attacked by sodium carbonate (trona), which causes severe cracking to occur along joints and other lines of weakness) (photograph by A. S. Goudie)

are normal marine brines, but in parts of the inner lagoon the dissolution of gypsum from bottom sediments has produced brines with very high calcium and sulphate concentrations. Along the southern parts of Bardawil lagoon there are belts of sabkhas, inundated occasionally by lagoon waters. These sabkhas contain normal marine brines, as well as calcium chloride brines formed where the marine brines react with sediments rich in calcium carbonate. Gavish (1980) provides a useful review of the different types of coastal sabkha environment, with special reference to the southern Red Sea coasts of the Sinai peninsula, and shows how "evaporative pumping" of groundwater causes different minerals to precipitate out at different levels. Carbonates are precipitated first, with sulphates above and halites forming a crust within the higher, supratidal zone.

The most important sabkhas are those that border the western shore of the Arabian Gulf (Figure 2.10). Their geochemistry has been the subject of lengthy study (Warren, 1989) and is of significance because of the large amount of construction activity that has taken place in their vicinity in recent decades. The Arabian Gulf sabkhas are unique because of the large volumes of nodular anhydrite (calcium sulphate) currently forming in the intertidal zone. The anhydrite develops because of the very high temperatures and salinities characteristic of this environment; southern Kuwait appears to be the northern limit of contemporaneous anhydrite formation. Gypsum is also common in the sabkha sediments, but there are no significant concentrations of halite, largely due to the high humidity of the area. Much of the sodium chloride which is deposited is either blown inland by winds or is redissolved by flood waters and returned in the subsurface seaward flow of groundwaters. A good review is provided by Evans (1995).

Evaporites in caves

Evaporite minerals are recorded from many caves and have been discussed in detail by White (1976: 304–316). They include gypsum, epsomite, thenardite, mirabilite, halite and various nitrates and phosphates derived from the urine, excrement and dead bodies of cave organisms. They play a part in cave development, as has been made clear by Jennings (1985: 32–33) in the context of Eocene limestones in Australia

> Some caves or parts of them are dry enough to allow crystallisation of salts by evaporation and so salt wedging can become important. Gypsum is the more widespread mineral to form and act in this way, but its effects are small in comparison with those of halite in Nullarbor Caves . . . , roof dome formation here may owe much to this process, and the products pile up in dune-like forms.

Gillieson (1996) reports that some of the speleothem decoration in Nullarbor caves is broken down by salt weathering.

Another location where salt attack has been shown to be important in cave development is in Kentucky, USA. The mechanism in that example was explained thus in White (1988: 235)

> The gypsum and other sulfate minerals in the Mammoth Cave area originate from the oxidation of pyrite in the upper part of the overlying Big Clifty sandstone (Pohl and White, 1965). Sulfate-bearing solutions percolate downward into the cave passages, where they react in the limestone and form gypsum by direct in situ replacement of calcite. The molar volume of gypsum is higher than that of the calcite it replaces, thus producing an expansion pressure in addition to the chemical attack. Replacement within the bedrock causes thin, platy fragments of rock to spall off, resulting in a characteristic chip breakdown. Expansion within the beds and slippage along the softer gypsum stringers also generate a peculiar curved slab breakdown.

Tunnels made by humans may also develop aggressive efflorescences, as was noted by St John (1982) in the context of railway tunnels in New Zealand, where sodium sulphate precipitated as a result of the evaporation of groundwater caused by the high levels of ventilation in the tunnels.

According to Ford and Williams (1989) gypsum, plus the minor but frequent epsomite and mirabilite, are the most common evaporite minerals in temperate to tropical zone caves that are seasonally dry. They form less commonly in cold, humid caves and in arctic caves where temperatures are -1 to $-3°C$. Other sulphate and halide salts are found only in warm, dry caves. Phosphate and nitrate minerals are largely produced from bat guano and other animal matter in caves reacting with rocks and speleothems.

Table 3.5.   Some influences on salt chemistry

Inputs
        Sea water
        Dust
        Rainfall
        Volcanic emissions
        Solutions derived from weathering of bedrock and ancient evaporities
        Submarine gas emissions
Changes within basin
        Cationic exchanges with clays
        Organic influences (e.g. algae, bacteria)
        Temperature changes
        Chemical reactions
        Tectonic changes
        Vertical leaching processes:
          (i) Affected by climate
          (ii) Topographic position
        Deflational removal
        Evaporative concentration
        Capillary migration

Nitrates have been extracted from cave deposits for centuries for use in medicines and ceramics.

Controls on the nature of salt type

As Table 3.5 indicates, there is a wide range of factors that can account for the observed variety of salt lake and efflorescence types described in this section. These factors can be divided into two principal groups: those involving the nature of inputs into the water–salt system and those involving changes within the system. The nature of salts, both in lakes and efflorescences, depends in part on the nature of the materials that come into a particular area. A good review is provided by Smoot and Lowenstein (1991).

One obvious source of saline material is sea water. In areas that were originally occupied by the sea, regressions, tectonic changes or geomorphological processes (e.g. spit and bar growth) may lead to the isolation of bodies of saline water that have many of the chemical characteristics of sea water. This is the "relict sea water theory" of Johnson (1980). For example, Godbole (1972) interprets the high sodium chloride level of the salt lakes of Rajasthan (India) as the result of the retreat of the Tethys Ocean, while Busson and Perthuisot (1977) regard the sabkhas of South Tunisia, such as the Sebkha el Melah, as being erstwhile coastal areas that have become isolated from the Mediterranean by a combination of Holocene tectonic subsidence and barrier growth. These types of basin are termed "paralic".

Figure 3.4. In coastal situations, spray derived from the ocean can cause the accumulation of substantial amounts of salts, which can then cause aggressive weathering. This picture shows salty spray being deposited on volcanic cliffs on the west coast of Fuerteventura, Canary Islands (photograph by A. S. Goudie)

Another major source of inputs are those derived from the atmosphere (Figure 3.4), either as aerosols (Zezza and Macri, 1995) or as rainfall, aeolian dust, volcanic vapours or submarine gas emissions. This is the "cyclic salt theory" of Johnson (1980). The first of these, rainfall, has often been proposed as the source of the high chloride contents of groundwaters in Israel and of the salt lakes of Rajasthan and Australia (see Eriksson, 1958; and Table 3.6). Rainfall chemistry varies with distance from the sea and in inland situations the importance of chloride is reduced relative to sulphates and carbonates. Near coasts, however, sea salt can be a major control of rainwater chemistry. Eriksson (1960), for example, estimates that chlorine deposition on exposed western coasts in Europe can reach $10–20\,g\,m^{-2}\,yr^{-1}$, and Shaw (1991) found that sea salt particles can be carried some $900\,km$ inland. Dry deposition of marine aerosols can also occur near the coastal zone, as shown by the detailed monitoring of Gustafsson and Franzen (1996) along the coast of south-west Sweden, where westerly gales blow salt several kilometres inland. The same relationship has been found for efflorescences in Antarctica (Keys and Williams, 1981). The cyclic salt theory has now been dismissed, however, as an adequate explanation of the sodium and chloride dominance seen in Australian lakes (Johnson, 1980: 236)

> The theory does not explain, for example, why there are some clay pans and salt pans, freshwater lakes and salt lakes adjacent to each other both on the coast and

Table 3.6.   Rates of salt addition to arid areas from precipitation. From data in Eriksson (1958). Reproduced by permission of Unesco

Area	Deposition of chloride (kg ha^{-1} yr^{-1})	Deposition of all sea salts (kg ha^{-1} yr^{-1})
Mexico	2	—
Patagonia	1	—
Sahara steppe	3	—
Namib	80	120
Bloemfontein, S. Africa	—	5
Kalahari	—	2
Rajasthan, India	2	5.5
Interior Australia	5	—
Near East	1	—
Gobi Desert	—	0.5

inland. Nor does it explain why the waters along the coastline of the continent are not all saline . . . or why more of the continent is not covered in sea salt if it is brought to all areas via the same medium.

However, Jones et al (1994) have used salt balance and isotopic signatures to investigate solute sources in the central Murray Basin, Australia and found them to be marine in character. They suggest that over the last 0.5 million years significant quantities of marine salt have been blown in as aerosols and have subsequently leached into shallow regional groundwaters.

Related to inputs in rainfall is that in fogs. In the coastal zone of the Central Namib, fog water can account for the deposition of as much as 100–1000 kg ha^{-1} of sodium chloride deposited each year (Walter, 1937; Boss, 1941). Some more recent data on fog water chemistry are provided by Schemenauer and Cereceda (1992a and b) for Oman and Chile (Table 3.7). Their data indicate that although the concentrations of major ions are not especially high, the fog water does have a significant content of some ions which provide a source of salts on rock surfaces. Table 3.8 gives an analysis of a fog from the Atacama Desert; the significant quantity of nitrate it contains may, over an extended period of perhaps 10–15 million years, explain the development of the Chilean and Peruvian nitrate deposits (Eriksen, 1981: 32). However, it is unlikely that this is the only source of the Atacama nitrates; volcanogenic atmospheric fallout and groundwater of Andean origins are also probably important (Searl and Rankin, 1993).

Dust derived from the deflation of desert playas, coastal sabkhas and other susceptible surfaces may also provide inputs of saline materials to closed basins and onto desert rock outcrops. For example, Bucher and Lucas (1975) found that carbonates amounted to 20–30% in Saharan dust deposited in the Pyrenees; Yaalon and Ginzbourg (1966) reported calcium carbonate contents

Table 3.7. Comparison of the concentrations in mg l^{-1} of major ions in fog-water samples from coastal Oman and Chile

Species	Oman*	Chile[†]
pH	7.0–7.9	3.5–6.7
SO_4^{2-}	3.4	9.1
NO_3^-	4.7	2.0
Cl^-	44	7.9
F^-	0.02	—
HCO_3^-	10.8	—
Na^+	24	4.8
NH_4^+	0.2	1.2
K^+	1.1	0.3
Ca^{2+}	15	1.0
Mg^{2+}	2.9	0.6

*Schemenauer and Cereceda (1992a).
[†]Schemenauer and Cereceda (1992b). Both reproduced by permission of the American Meteorological Society.

in Negev dust of up to 48% and soluble salts up to 3.1%; Warn and Cox (1951) found that at Lubbock, Texas, carbonate equalled 5–20%, and gypsum 5%; whereas on the Canary Islands Logan (1974) found that soluble salts amounted to 1.2–3.6%. In Tunisia, dust deflated from the great inland sabkhas has been shown to make a major contribution to the development of the extensive gypsum crusts (Coque, 1961; Watson, 1983). On the Red Sea coast of Sudan, aeolian dust consists of aggregates cemented by halite (Schroeder, 1985). Large quantities of saline dust are also reported as being blown off the desiccating bed of the Aral Sea.

The most comprehensive survey of dust additions of saline materials to desert surfaces is that undertaken in the western USA by Reheis and co-workers (Reheis et al, 1995). Reheis and Kihl (1995) monitored the salt

Table 3.8. Chemical composition of a 'camanchaca' fog from the Atacama Desert. From Eriksen (1981)

Species	Concentration (mg l^{-1})
Cl^-	46
SO_4^{2-}	32
NO_3^-	19
HCO_3^-	15
Na^+	30
Ca^{2+}	12
Mg^{2+}	6
K^+	1
SiO_2	1

content of dust in southern Nevada and California from 1984 to 1989 and found that the average soluble salt content (excluding gypsum) ranged from 4 to 19%, equivalent to a salt flux of 0.3–2.4 g m^{-2} yr^{-1}. The gypsum content ranged from 0.1–7%, equivalent to a flux of 0.02–1.5 g m^{-2} yr^{-1}. Thus the total saline inputs from dust can approach 4 g m^{-2} yr^{-1}.

In the Namib, Martin (1963) suggested that some of the sulphate in the gypsum crusts was indirectly derived from hydrogen sulphide emissions from the offshore waters, while in the case of the sulphate-rich efflorescences of the Mount Erebus area of Antarctica and the nitrate beds of the Atacama there has been discussion about the possible contribution of volcanic gases (Eriksen, 1981; Jones et al, 1983).

Probably one of the most important controls on basin chemistry is the nature of the inputs from rock weathering. For example, whereas the waters of Lake Lahontan (USA) are distinctly alkaline and contain much sodium carbonate, those of the Bonneville basin are dominantly of sodium chloride. Clarke (1916: 159) attributed this difference to the fact that the Lahontan lakes are supplied with water from areas of igneous rocks, in which andesites and rhyolites are especially abundant, whereas the Great Salt Lake of the Bonneville basin is fed by streams and springs which flow mainly through sedimentary formations.

This basic distinction is one that has been followed a great deal since the work of Clarke (1916). Strakhov (1970: 276), for example, suggests that soda lakes are confined to one of two types of arid zones: where acid, intermediate or alkaline crystalline rocks are exposed (e.g. some of the East African Rift Valley lakes) or where there are polymictic sandy deposits rich in feldspars and other aluminosilicates (e.g. the basins of the Kulunda and Kurzan steppes in the former USSR). One possible explanation for the sodium chloride dominance of Australian waters is the lack of widespread recent volcanic activity (De Decker, 1983).

Where bedrock contains old evaporitic beds (as with the bedrock gypsum of Tunisia), leaching may transfer material to modern evaporite basins (e.g. the Chotts). This is the "connate salt theory" and has been put forward as the explanation of the high sodium and chloride levels in Australian salt lakes (Johnson, 1980: 259)

> Australia is a land of very subdued relief—in Australia's recent past there has been no major tectonic activity to create the mountain ranges from which significant rivers could emerge to flush the land free of the residue of transgressing seas. There the fossil sea salt of ancient marine rocks has not been carried away but merely redistributed by the wind or by slowly meandering rivers which never reached the sea.

Leaching of Palaeozoic evaporites may account for the sodium sulphate and magnesium sulphate brines of Saskatchewan (Last and Schweyen, 1983). Some shales may also contain appreciable quantities of salts. This applies to the famous nitrate shales from the Cretaceous of Egypt, which contain appreciable quantities of sodium nitrate, sodium sulphate and sodium chloride (Hume, 1925: 205–209).

Most rocks contain some primary content of salt or of chemical elements that can form salts after they have been released from the crystal lattice of rock-forming minerals by chemical weathering. Kirchner (1996: 88) gives the following mean values for chloride and sulphur derived from the literature:

Granites	130–240 ppm Cl
	270–440 ppm S
Basalts	50–160 ppm Cl
	250–300 ppm S
Metamorphic rocks	207–354 ppm Cl
Claystones and shales	100–180 ppm Cl
	2408–3000 ppm S
Carbonates	130–660 ppm Cl
	240 ppm S
Sandstones	10–20 ppm Cl
	240 ppm S

Climate also affects the distribution of salts. It is only in extremely arid areas that highly soluble salts are not leached from surface materials. This may, for example, be the explanation for the development of nitrate deposits in the Atacama desert. Sodium nitrate is highly soluble in water compared with most other common crust materials, and the Atacama is probably the driest of the world's deserts, with the average annual rainfall being less than 1 mm in the areas where the nitrate deposits are most prevalent.

Evaporation of brines occurs in dry climatic conditions and leads to a sequence of salts precipitating out of an increasingly concentrated solution. Chemists have long found difficulty in understanding the exact behaviour of such concentrated solutions because of the difficulty in computing activity coefficients at high ionic strengths.

Another important climatic effect is that of temperature (Thornton and Nelson, 1956), because the concentration of brines can take place by freezing. Sodium sulphate is highly susceptible to precipitation from saline waters and is precipitated around many saline lakes and in saline soils during the winter months.

The chemistry of lake solutes is also influenced by the chemical fractionation that takes place in closed basins between dilute inflow and concentrated brines (Eugster and Jones, 1979). Of the eight or so most abundant initial solutes (silica, sodium, potassium, calcium, magnesium, bicarbonate, sulphate), some are removed preferentially from solution while others increase in abundance. The mechanisms involved include mineral precipitation, the selective dissolution of efflorescent crusts and sediment coatings, absorption on active surfaces (e.g. on clays), degassing (resulting from equilibration with the atmosphere, an increase in temperature, a decrease in solubility with salinity, or organic activity such as photosynthesis) and redox reactions.

One major cause of mineral precipitation is the reduction in solubility of some salts by the presence of others, apparently as a result of the common ion effect. For example, as the content of sodium chloride in a brine increases by, for example, evaporative concentration, the solubility of both sodium carbonate and sodium sulphate decreases. The presence of 10% sodium chloride in a solution (Hunt et al, 1966) reduces the solubility of the carbonate and sulphate by one-third or one-half, and in brines as concentrated as 25% sodium chloride there may be small quantities of sodium sulphate but no sodium carbonate.

The reaction of brines with minerals within the basin may also change their nature. Levy (1977), for example, noted that calcium chloride brines and diagenetic dolomite were produced in Sinai sabkhas by the reaction of sea water with pre-existing carbonate minerals.

Biological influences on the precipitation of salts are also important, even in the highly inhospitable dry, saline environments just discussed. Within many sabkhas and saline lakes, for example, a productive, if not very diverse, ecosystem exists. Larsen (1980) reviewed the nature and structure of hypersaline ecosystems and stressed that "The indigenous population of hypersaline environments are, as a rule, rather special organisms adapted to live in the strong brine, and they even prefer, or require, the high salinity in their environment for growth and reproduction" (Larsen, 1980: 24). The Great Salt Lake, Utah, for example, contains populations of the brine shrimp *Artemia salina*, halophytic, chemo-organotrophic bacteria from the genera *Halobacterium* and *Halococcus* (which are red in colour) as well as the brine algae *Dunaliella salina* and *D. viridus*. The halophytic bacteria require minimum concentrations of 10–15% sodium chloride to survive, and flourish best in almost saturated salt solutions. Algal mats, composed of a range of algae and cyanobacteria, are found along many sabkha shorelines and may play a key part in chemical reactions and the precipitation of minerals. Gavish (1980) reported that at the Ras Muhammed pool in the southern Sinai peninsula, algal mats up to 30 cm thick grow on the bottom of the pool, with layers of gypsum crystals and carbonate mud underneath. On rock surfaces within many hot and cold desert areas a rich microflora containing lichens, algae, bacteria and fungi develops (e.g. as found by Danin and Garty, 1983, in the Negev desert), which may influence the weathering action of salts. Rich associations of micro-organisms forming biofilms are also present on many building material surfaces and may play a part in salt weathering in such settings. Strzelczyk (1981) provides a good review of the role of stone-dwelling bacteria in the oxidation of sulphur and ammonia to form nitrate salts.

In many situations, the chemistry and mineralogy of salt deposits and efflorescences will depend on a combination of the factors we have just discussed. A number of sources and pathways will be involved. This is illustrated in the context of Antarctica in Figure 3.5.

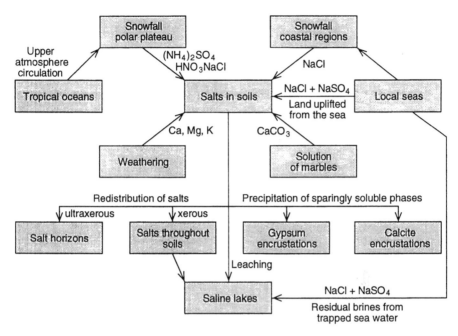

Figure 3.5. Flow chart showing the probable pathways for derivations of the salts in Antarctic soils. Modified from Campbell and Claridge (1987: figure 9.11). Reproduced with permission of Elsevier Science

Efflorescences on walls

In recent years an increasing amount of attention has been given to the nature of efflorescences on walls of buildings and engineering structures. Table 3.9 gives some data on the range of efflorescences that has been recorded. Even though the range is substantial, gypsum is probably the most common efflorescence that has been found on building materials, though sodium sulphate, sodium chloride and sodium carbonate may also be relatively common. One of the most comprehensive national surveys of wall efflorescence types is that undertaken in Sweden by Nord (1992). Rapid methods for the analysis of such efflorescences are discussed by Arnold (1984). Studies using fresh natural stone samples as sensors reveal the short-term development of efflorescences on building stones. Thus, for example, Viles (1990) reported the development of small gypsum crystal growths on Portland limestone samples exposed for only two months in central London (Figure 3.6). The National Materials Exposure Programme in Britain exposed small samples of various stone types at 29 sites under rain-sheltered and exposed conditions for a period of at least four years. In most cases 70–90% of the

Table 3.9.   Efflorescences on buildings

Location	Salts	Reference
Sphinx, Egypt	Gypsum, halite	Helmi (1990)
Church Frescoes, Germany	Gypsum, sodium sulphate	Rösch and Schwarz (1993)
Venetian fortress, Corfu	Halite	Moropoulou et al (1993)
Konya Mescid, Turkey	Potasium nitrate, gypsum, thenardite	Tuncoku et al (1993)
Mamluk and Turkish buildings in Cairo, Egypt	Halite, gypsum and calcite	Abd el Hady (1992)
Almeric Cathedral, Spain	Dolomite, calcite, gypsum, hexahydrite, grunerite, tremolite, humberstonite, epsomite	Martin et al (1992)
Karnak temples, Egypt	Gypsum and thenardite	Bromblet (1993)
Cadiz Cathedral, Spain	Halite, sodium sulphate, sodium carbonate	Galan (1990)
Elgin Cathedral, UK	Calcium sulphate	Laurie (1925)
Durham Cathedral, UK	Calcium sulphate, magnesium sulphate	Laurie (1925)
Westminster Hall, London	Calcium sulphate	Laurie (1925)
Eglise Notre Dame La Grande, Poitiers, France	Calcium sulphate and halite	Hammecker and Jeannette (1988)
St Mark's Basilica Venice, Italy	Calcium sulphate	Fassina (1988)
Lincoln Cathedral, UK	Gypsum	Butlin et al (1988)
Swiss churches	Natrite, nitrokalite, mirabilite, epsomite, nitronatrine, nitromagnesite	Arnold and Zehnder (1988)
Rhodes Castle, Greece	Halite and sodium sulphate	Theolakis and Moropolou (1988)
Khiva, Uzbekistan	Calcite, halite and saltpetre	Cooke (1994)
Bukhara, Uzbekistan	Hexahydrate, epsomite, thenardite, saltpetre, nitratine, halite and gypsum	Cooke (1994)

Continued

Table 3.9. *continued*

Location	Salts	Reference
Petra, Jordan	Halite, gypsum, potassium nitrate, magnesium sulphate	Fitzner and Heinrichs (1994)
Malaga Cathedral, Spain	Magnesium sulphate, gypsum, halite, kalicinite, nahcolite	Carretero and Galan (1996)
Almeria Cathedral, Spain	Hexahydrite and epsomite	Villegas Sánchez et al (1996)
S. Maria dei Miracoli Church, Venice, Italy	Aphthitalite, thenardite, gaylussite and trona	Fassina et al (1996)
Oporto and Braga, Portugal	Gypsum and halite	Begonha et al (1996)
Freiberg, Saxony, Germany	Epsomite, hexahydrite and starkeyite (leonhardtite)	Klemm and Siedel (1996)
St Maria im Kapital, Cologne, Germany	Halite, nitratine, thenardite mirabilite, epsomite, gypsum and trona	Laue et al (1996)

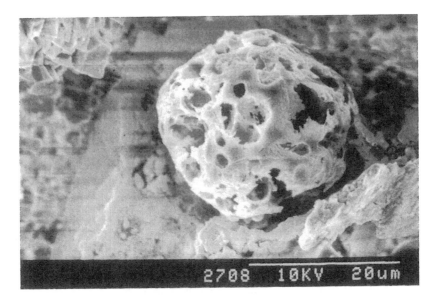

Figure 3.6. A carbonaceous pollution particle, probably from oil-fired combustion, which was deposited on test Portland Stone substrates exposed in central London for two months. Gypsum also developed on the surface over the two-month period (photograph by H. A. Viles)

measured weight gain of sheltered samples was accounted for by soluble salts, dominantly gypsum, but with soluble nitrates and sodium chloride also found (Butlin et al, 1992).

The nature of efflorescences on walls depends on many of the factors we have already discussed with respect to efflorescences on natural rock outcrops or with respect to lake salts. However, in the case of walls there are additional factors that need to be considered which can be split into two main types: those relating to air pollution and those relating to building materials and practices. Looking firstly at air pollution, nitrogen oxides and particularly sulphur dioxide have been implicated in various chemical reactions at the surface of buildings which can produce salts and ongoing salt weathering damage. A range of other pollutants may be involved, such as soot particles. While most attention has been directed to the external walls of buildings, some decay induced by air pollution can also take place on internal walls. Laurie (1925: 89), for example, attributes some of the internal decay of Ely Cathedral in eastern England to the fact that it was "occasionally lighted by gas, and is heated by pernicious stoves".

Sulphur dioxide reacts with building materials containing calcium carbonate through the process of sulphation (as described more fully in Chapter 5) to form gypsum, which is by far the most common efflorescence found on building stones. Simply put, this involves the following reaction under conditions of >80% relative humidity:

$$CaCO_3 + H_2SO_4 + H_2O \rightarrow CaSO_4 \cdot 2H_2O + CO_2$$

where the H_2SO_4 is produced by the oxidation of sulphur dioxide. Catalysts for this reaction are found in soot and dust, and atmospheric oxidants such as ozone and hydrogen peroxide are also important. Fly ash has also been suggested to be an important contributor of calcium and sulphate ions and a catalyst for the sulphation process (del Monte and Vittori, 1985). However, experimental studies using Portland and Monk's Park limestones have not demonstrated any major catalytic role of fly ash as iron-containing compounds within these limestones are sufficient to catalyse the necessary oxidation of SO_2 to SO_4^{2-} (Hutchinson et al, 1992). Simulations in the Lausanne Atmospheric Simulation Chamber (Ausset et al, 1996) using different carbonate stone types have reasserted the role of fly ash. Thus the debate continues.

The catalytic role of carbon (soot) and metal-rich particles from vehicle exhaust has been studied by Rodriguez-Navarro and Sebastian (1996). Their work on limestone from Granada Cathedral in Spain and experimental studies confirm such a catalytic role. The sulphation reaction is encouraged by the presence of moisture on the stone surface and within pores. The resultant gypsum is more brittle and more soluble than the stone from which it is formed; where runoff occurs frequently the soluble gypsum is washed away, and where it is allowed to accumulate it can cause much damage to the underlying stone.

Gypsum occurs predominatly in three main settings on building façades. Firstly, as a major component of black crusts (comprised of organic and inorganic airborne particles, inorganic precipitates and organic growths). Secondly, in areas where water and soluble salts are allowed to percolate through walls, in which case the gypsum largely originates from building materials such as mortars. Finally, gypsum may occur within the zone of rising damp, where it precipitates out both early on at low levels and in the upper fringes of the zone (Zehnder, 1993). Two types of gypsum-enriched black crusts can be identified: those forming on stone which does not contain much calcium and therefore where the crust simply "sits" on the surface, and those forming on calcium-rich stone where the gypsum is largely a reaction product with the underlying stone.

Nitrogen oxides may be important aids to salt weathering in two main ways. Firstly, nitrogen dioxide enhances the adsorption of sulphur dioxide by the stone surface (Haneef et al, 1992). Secondly, nitrogen oxides may be oxidised in the presence of water to produce nitric acid, which may react with carbonate stone to produce highly soluble calcium nitrate salts. As yet there has been no firm evidence that such nitric acid-induced salts are retained on or within stones, or contribute significantly to decay. However, as nitrogen oxides become an increasingly important element of urban air pollution in many parts of the world, their role in stone decay needs to be clarified.

Air pollution may also react with natural sources of salts to produce particularly damaging environments, as found in some coastal zones. Detailed studies of fog water and interstitial aerosol chemistry were made in Berkeley, California (Gundel et al, 1994) during 15 fog episodes in the summer of 1986 which show such reinforcing influences. Na^+ and Cl^- were the most abundant ions in the fogs, followed by NH_4^+, SO_4^{2-} and NO_3^-. Based on an equivalence ratio of sulphate and sodium ions of 0.119 in sea water, an average of 18% of the sulphate in the fog water originated from sea salt, with pollution presumed to account for much of the rest. The sample site was 27 km inland and the fogs passed over polluted urban areas before reaching the site. The fogs had a mean pH of 4.0. Gundel et al (1994) concluded that the sulphate which was not derived from sea salt had been formed in the fog.

The effectiveness of disintegration in walls induced by gypsum can be heightened by what is often called the limestone over sandstone effect (Livingston, 1994: 103). It has frequently been observed that sandstone courses lying beneath limestone courses suffer from especially severe attack. Schaffer (1932: 2) gives many examples from Bristol and elsewhere in the UK, while Arkell (1947: 156–157) noted that in Oxford, where red sandstone had been used as a decoration in limestone buildings (e.g. Balliol College chapel), that this "not only causes offence to the aesthetic sense of the present generation but will be a drain on the resources of a future generation". The reason given for this is that gypsum is formed on the overlying limestone though its reaction with sulphur dioxide derived from the atmosphere or from acid precipitation. The gypsum is then leached from the limestone by rainwater which flows into the sandstone (which may be quite porous). There the water evaporates, leaving behind the gypsum which can recrystallise and cause weathering of the sandstone course.

A similar effect can occur if limestone is introduced into a magnesian limestone building, but in this case it is magnesium sulphate that causes the high rates of decay (Schaffer, 1932: 22).

The relatively common occurrence of sodium sulphate on urban walls may also be explained, at least in part, by air pollution, which provides the sulphur. The sodium may originate from diverse sources including groundwater, the stone itself, de-icing salt or sodium-containing liquids used to clean façades or to remove old paint (Nord, 1992: 425).

The permeable nature of many pointing and bedding mortars suggests that they are subject to high levels of water movement by wetting, absorption and evaporation. Consequently, they may retain and/or release salts which are damaging to building stones. Indeed, stone around mortar joints may often be observed to be decaying for this reason.

Salts may be produced by reactions involving paints (see, for example, Garcia-Vallès et al, 1996) or mortar, a process which Laurie (1925: 86) termed "infection". For example, the lime mortar of old buildings may be made of dolomitic lime, the dolomitic compounds of which (see, for example Villegas Sánchez et al, 1996) may react with sulphate ions brought in by water and from the air as follows (Arnold and Zehnder, 1990: 32):

$$CaMg(CO_3)_2 + SO_4 \rightarrow CaCO_3 + MgSO_4 + CO_3$$

Salts may also issue from Portland cement (see, for example, Fassina et al, 1996). This may contain up to 1% of soluble alkalis. The ions leached out may form efflorescences of alkali carbonate salts including trona (Charola and Lewin, 1979). Arnold and Zehnder (1990: 33) provide an indication of the quantities that may be produced

> The amounts of portland cement used in walls being very large the quantities of soluble salts may become very important too. As an example, 100 kg of portland cement with a content of 0.1% of soluble Na_2O may produce 460 g of natrite ($Na_2CO_3 \cdot 10H_2O$) or when reacting with sulphuric acid from the polluted air 520 g of mirabilite ($Na_2SO_4 \cdot 10H_2O$).

A recent study of granite weathering on façades of Trinity College, Dublin exemplifies the often complex influences of mortars and other building materials (O'Brien et al, 1995). Initial observations showed a clear correlation between areas of granite showing serious decay and the accumulation of calcium salts. Detailed microscopic observations revealed that salt-affected areas were heavily fractured and oxidised, with the delamination of mica, the alteration of feldspars and chlorite and blackening by carbonaceous particles all common. Salt action seems to be enlarging such fractures, thus permitting water and soluble salts to permeate the stone, producing yet further decay. Portland limestone (used for decorative purposes on the façades) and calcium-rich mortars are seen to be the main contributors of soluble salts here, and the mortars furthermore act to concentrate damaging water attack, especially when their strength, flexibility and permeability differ markedly from the surrounding stone.

Another influence on the nature of efflorescence on walls is the pattern of water movement in the walls themselves. Transported salts are precipitated in a spatial sequence (see, for example, Hammecker and Jeannette, 1988) according to the ion activities of the salt phases in the system. In the lower zone less soluble and less hygroscopic sulphates and carbonates are mainly present, while in the upper zones chlorides and nitrates forming very hygroscopic solutions tend to accumulate (Arnold and Zehnder, 1990: 37–8).

Soluble sulphates are present in many fired clay bricks and can be redistributed to give efflorescence. It is possible to obtain special low sulphate bricks that have maximum permissible amounts of sulphate as laid down in British Standard BS 3921 (Addleson and Rice, 1991) to avoid this problem.

The presence of salts in clay bricks comes from three main sources: the clay itself (including the tempering water, pyrites and the action of sulphur from the fuels used in firing; Addleson and Rice, 1991: 336). Calcium and sulphate are the dominant ions in the soluble salt component of British bricks and the total soluble salt content (as weight per cent) ranges from 0.2 to 5.7%. The temperature of firing also tends to determine the salt content because certain salts may be decomposed and expelled from bricks by hard firing.

Colliery shale waste is another source of sulphates, and may contain 1–5% of soluble salt content. If concrete floors are laid above it and appropriate precautions are not taken, these soluble salts may migrate into the concrete and cause deterioration. Pyretic shales are especially prone to produce gypsum, which can cause the heaving of floors in buildings (Hawkins and Pinches, 1987). It is even possible that gypsum, implicated in the decay of Lincoln Cathedral in England (Figure 3.7), may not all be derived from atmospheric pollution. The Lincolnshire Limestone, when freshly excavated, has been seen to contain ooids, the cortices of which are packed with crystals of iron sulphide. This could, after oxidation and other mineralogical changes, provide a source of sulphate in the rock mass (Jefferson, 1993).

Some efflorescences on walls can be produced by what Arnold and Zehnder (1990: 35) term "biologic metabolisms". In some cases humans and animals deposit excrement or urine, which can contain appreciable quantities of chlorides and nitrates (Figure 3.8). In addition, however, bacteria can oxidise ammonia to produce nitrate salts, while lichens living on calcareous rocks may secrete oxalic acid, which reacts with calcite to form calcium oxalate.

Figure 3.7. The West Front of Lincoln cathedral showing blackening of sheltered surfaces and the wooden boxes constructed as temporary measures to protect the medieval romanesque friezes in the 1980s (photograph by H. A. Viles)

Figure 3.8. A decaying farm building wall, near Chelmsford, Essex, UK. The efflorescence, developed on brickwork, has caused alveoles to form. The source of the salt is believed to be animal waste that has accumulated in the animal stalls within the building (photograph by A. S. Goudie)

Several studies have been made of oxalate crusts and efflorescences on building surfaces which support the hypothesis that they have been produced by lichens. Del Monte (1991), for example, studied the unevenly distributed brownish patina (locally called *scialbatura*) on Trajan's Column in Rome and found that it consisted mainly of a mixture of whewellite and weddelite (mono- and dihydrate calcium oxalates) and small quantities of gypsum. Del Monte ascribes the oxalates to epilithic and endolithic crustose lichens whose chemical exudates have reacted with the underlying Carrara marble. Other studies, for example, by Vendrell-Saz et al (1996), have shown that patinas from a wide range of buildings in the Mediterranean area contain calcium carbonate, oxalates, phosphates and gypsum, all of which may be at least partly a result of biofilm activity producing microstromatolites through a range of biomineralisation processes.

Finally, Zehnder and Arnold (1984) have attributed damage to the baroque "Erlacherhof" in Berne, Switzerland to the formates of calcium and magnesium produced after formic acid had been used as a cleansing agent on the sandstone structure.

SOURCES OF MOISTURE FOR SALT WEATHERING

Besides the availability of salts, another crucial control of the effectiveness of salt weathering is the presence of sources of moisture that can either provide a source of salt or a means by which salts can be mobilised and deposited in pores and cracks.

The main sources that merit consideration include dew, fog, rain and groundwater.

Dew Frequency

On rock outcrops in some desert areas, appreciable amounts of dewfall may occur on a suprisingly large number of nights in the year and may therefore provide conditions for salts to hydrate and for hygroscopic salts to take up moisture. For example, Ashbel's (1949) pioneering work in Palestine demonstrated that some stations received dew on over 200 occasions in the year and that amounts of dewfall could exceed 100 mm per year (Figure 3.9, Table 3.10).

More recent observations of dews in the Negev Desert have been made by Zangvil (1996) based on observations at Sede Boker. The total amount of moisture deposited each year averages about 17 mm and the number of dew hours per night averages between around 2.5 and 5 hours, depending on the season. The number of dew events ranges between about 8 and 26 days per month, with minimum numbers in April and maximum numbers in September.

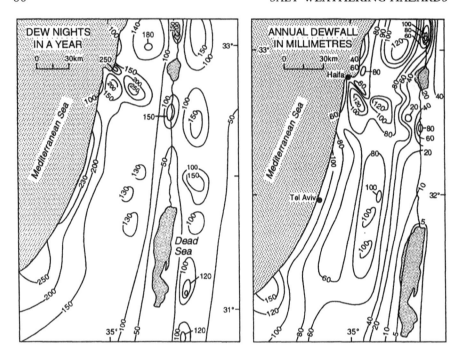

Figure 3.9.   Dew nights and dewfall in Palestine. Modified after Ashbel (1949: figures 1 and 2)

Fog Water

Fog may contribute to salt weathering both by adding appreciable quantities of moisture to rock surfaces and by being a potential source of salts. Fog water is of special importance in coastal deserts such as the Atacama and Namib, and the latter area will be used here to illustrate the frequency with which fog occurs in such environments and to ascertain how much moisture is precipitated from fog. The data are all derived from Lancaster et al (1984).

In the Central Namib (Figure 3.10) there are nine stations with useful fog data. As a result of moist oceanic air flowing over the upwelled cold Benguela current, the effects of fog are felt more than 100 km inland. At a number of stations *precipitating* fog occurs on between about 50 and 90 days in the year, though the number of days when fog is observed may be greater than this and figures as great as 300 days in the year have been given. Fog water precipitation exceeds the mean annual rainfall and increases from the coast inland to a distance of about 35–60 km from the sea and then decreases further inland. Annual fog precipitation values at two inselberg stations exceed 180 mm per year (compared with around 20 mm of annual rainfall). Thus the mean amount of fog precipitated per foggy day or night can be several millimetres (Table 3.11).

Table 3.10. Number of dew nights for selected stations in Palestine

Station	Lat N	Long E	J	F	M	A	M	J	J	A	S	O	N	D	Total for year
Gevulote	31°13'	34°28'	17	13	15	18	17	22	22	22	22	18	18	17	218
Beth Eshel	31°15'	34°59'	9	11	13	12	11	13	10	12	11	5	14	14	135
Dorot	31°31'	34°39'	16	16	14	15	14	11	11	15	17	18	17	20	182
Negba	31°40'	34°41'	8	6	6	8	13	10	12	12	15	15	17	12	134
Sarafand	31°57'	34°51'	11	8	8	14	19	22	25	26	24	15	16	11	200
Kefar Masaryk	32°54'	35°07'	12	15	18	22	26	27	25	26	26	23	20	15	255
Nahalal	32°41'	35°11'	15	13	21	20	22	27	29	26	24	18	15	12	242
Carmel	32°48'	34°59'	6	4	12	15	17	25	25	25	21	9	11	6	176
Kefar Etsyon	31°39'	35°07'	12	7	11	13	7	12	14	12	15	13	8	7	131
Jerusalem	31°47'	35°12'	11	8	6	6	6	10	10	13	14	12	8	9	113
Tirat Tsevi	32°25'	35°31'	13	13	14	13	15	15	14	13	12	11	13	16	162
Dan	33°14'	35°39'	14	13	20	21	20	22	27	27	29	20	11	16	210
Amman	31°57'	35°57'	14	8	8	4	2	1	2	2	4	6	4	7	70

Source: Ashbel (1949: table 1).

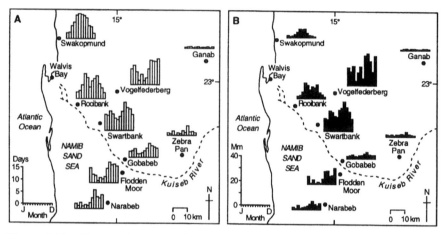

Figure 3.10. Fog conditions in the central Namib desert of Namibia. (A) Mean monthly number of days with fog-water precipitation. (B) Mean monthly fog-water precipitation values. Modified from Lancaster et al (1984: figures 9 and 10). Reproduced with permission

Rain

A further source of moisture for salt mobilisation and translocation in arid areas is rainfall. An important point to make here is that it rains more often in deserts than is commonly perceived and that rainfall events are not invariability of high magnitude and low frequency. As Table 3.12 demonstrates, the average rainfall amount per rainy day is quite modest, averaging just over 6 mm, with rather higher values than this for deserts in lower latitudes (e.g. west Africa and the Kalahari), but rather lower values for those in higher latitudes.

Groundwater

Groundwater is a major source of salts in suitable topographic situations, and the presence of high groundwater levels has been one of the prime determinants of the efficacy of salt weathering of engineering structures in the sabkha zones of the Middle East. As Cooke et al (1982: 168) point out

Aggressive soil conditions as an engineering hazard are produced when saline groundwaters are drawn upward through the soil to produce a capillary fringe that either reaches the ground surface or approaches sufficiently close to it to affect foundations. This upward movement of capillary water through surface and near-surface materials is sometimes known as "evaporative pumping" and is essentially the product of high surface temperatures. The height of capillary rise (i.e. the thickness of the capillary fringe), varies with the soil temperature gradient and the nature of the soil materials, normally being less than a metre in clean gravel but, according to laboratory studies . . . extends up to 10 m in clay.

Table 3.11. Fog data for the Central Namib Desert

Station	Mean annual fog precipitation (mm)	Mean annual number of days with precipitating fogs	Mean quantity of precipitation per foggy day (mm)
Flodden Moor	65.13	55.50	1.17
Ganab	2.67	2.76	0.97
Gobabeb	30.79	37.23	0.83
Narabeb	35.91	38.45	0.93
Rooibank	80.19	75.64	1.06
Swakopmund	33.94	64.68	0.52
Swartbank	183.62	86.71	2.12
Vogelfederberg	183.48	77.37	2.37
Zebra Pan	15.11	16.00	0.94

Terzaghi and Peck (1948) suggested that the height of capillary rise (Hc) that might take place in different materials could be estimated by using the formula

$$Hc = \frac{C}{eD_{10}}$$

where C = an empirical constant that depends on the grain shape and surface impurities, and ranges between 0.1 and 0.5; E = the void ratio; and D_{10} = "effective size", in which 10% of particles in the grain size analysis are finer and 90% are coarser than this value.

Cooke et al's own experiences in Bahrain, Dubai and Egypt suggested to them that capillary rise under desert conditions is normally no more than 2–3 m, and rarely, if ever, exceeds 4 m.

Table 3.12. Rainfall per rainy day in arid areas

Area	Number of stations in arid zones	Range of mean annual rainfall (mm) (30 year average)	Range of number of rainy days > 0.1 mm per year	Average rainfall per rainy day over 30 years (mm)
Former USSR	12	92–273	42–125	2.56
China	6	84–396	33–78	4.51
Argentina	11	51–542	6–155	5.41
North Africa	18	1–286	1–57	3.82
West Africa	20	17–689	2–67	9.75
Kalahari	10	147–592	19–68	9.55
Total	77	1–689	1–155	6.19

Capillary rise may take place in buildings and the height of such a rise, by what is sometimes termed "the wick effect", can be determined with a simple moisture meter, as was carried out in Ras Al Khaimah by Cooke and Goudie (1992, unpublished data), where walls built as recently as the late 1980s show clear evidence of severe salt attack. Depending on the nature of the substratum, moisture rises up breeze-block walls by as much as 1.8 m (Figures 3.11 and 3.12).

The position of the upper surface of the capillary fringe (known as "the limit of capillary fringe") with respect to the ground surface is of considerable importance in terms of building construction and land use planning and zoning. Cooke et al (1982: 169–170) believe that of particular importance is the line where the limit of capillary fringe intersects the ground surface, as this line marks the boundary between the zone where only foundations are at risk and that where both foundations and above-ground structures may be subjected to salt attack. They identified four zones (Figure 3.13) with different orders of susceptibility to hazard.

1. Zone I — no hazard from groundwaters as the limit of capillary fringe is so deep within the ground as to be below the base of foundations.
2. Zone II — the limit of capillary fringe is below the ground surface, but sufficiently close to it to affect foundations.
3. Zone III — the limit of capillary fringe is potentially above ground level so that both foundations and superstructures are at risk. Such areas usually show up as a dark tone on air photographs and have "puffy" surfaces on which footprints and tyre-tracks are clearly visible.
4. Zone IV — water-table within half a metre of the ground surface for most of the year so that the foundations are emplaced in water and there is the potential for capillary rise to well above ground level (sometimes in excess of 2 m). Such areas are very low in elevation and may be periodically inundated as a consequence of rainfall or sea surges. These areas appear as a dark tone on air photographs, often with patches of standing water. On the ground, the surface is usually either "puffy" or crusted. Patches of salt efflorescences are often developed and, in certain instances, salt polygons may also be visible. Standing water areas are invariably rimmed with salt crystals.

In some buildings, however, the thickness of the capillary zone may be exceptionally and unexpectedly high, sometimes exceeding 6 m (Cooke, 1994). Cooke (1994: 200–201) reported a suggestion developed by R. A. Legg which might account for this phenomenon, which was observed in some of the decaying Islamic monuments of Uzbekistan (Figures 3.14 and 3.15)

There may be two zones of capillary rise, one immediately above the water table (as normal), the other at or above ground level, the two separated by a relatively

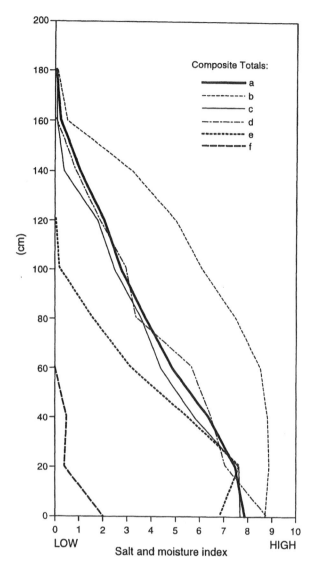

Figure 3.11. Salt and moisture profiles from walls from Ras Al Khaimah, United Arab Emirates. The scale on the x-axis represents the salt and moisture index determined by the protimeter and ranges from 0 to 10. Source: Cooke and Goudie (1992, unpublished data)

dry zone. The suggested mechanism for pumping water across the "dry zone" is as follows: in winter, when the surface is relatively cool and there is a marked temperature gradient from the water table to the surface, water will evaporate from the groundwater at depth, and the warm vapour will pass upwards through the dry zone to condense in the relatively cool surface layers. Such condensed

Figure 3.12. The "wick effect" demonstrated for a wall constructed on the salty sabkha surface in Ras Al Khaimah (photograph by A. S. Goudie)

water could be supplemented by rainfall, snowmelt, local runoff, irrigation water and floodwater. In the condensation zone, capillary rise could proceed into finer-grained material of the buildings from the surrounding soil.

FINGERPRINTING THE SOURCES OF SALTS

Stable isotopes (especially of sulphur) and ionic ratios can be used to identify the sources of salts in the environment, and thus help to explain the variety of

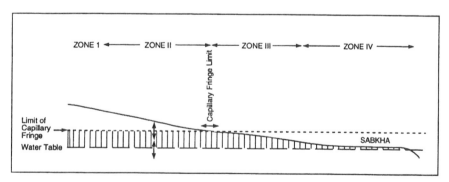

Figure 3.13. Hypothetical section showing zonal subdivision of aggressive ground in a sabkha environment. From Cooke et al (1982: figure v. 12). Reproduced by permission of Oxford University Press

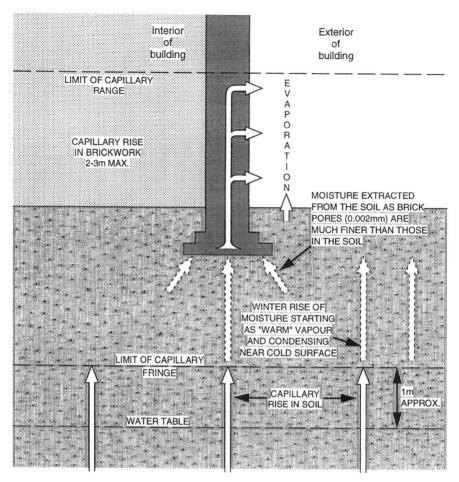

Figure 3.14. R. A. Legg's model explaining the capillary rise of moisture in buildings several metres above the water-table. Modified from Cooke (1994: figure 12.8) and reproduced with permission

salts involved in salt weathering. Chloride and sulphate can be particularly well "fingerprinted" in this way. Chloride is produced by evaporating salt spray droplets above oceans and pollution, whereas sulphate has a wider range of potential sources including sea spray, volatile biogenic sulphur (e.g. marine dimethyl sulphide), desert dust, marine gypsum and anhydrite as well as pollution sources (McCardle and Liss, 1995). Various studies have looked at the ratios of Na^+/Cl^- and Na^+/SO_4^{2-} in fog and rainfall—as well as those in salt efflorescences and brines—and compared them with the ratios found in sea water. Thus, as quoted earlier, Gundel et al (1994) used the ratio of

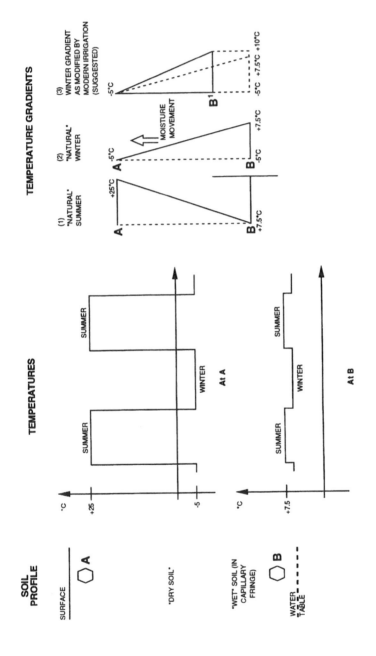

Figure 3.15. R. A. Legg's model explaining the double zone of capillary rise in terms of temperature gradients. Modified from Cooke (1994: figure 12.9) and reproduced with permission

SO_4^{2-}/Na^+ in fog water compared with that in sea water (0.119) to show the non-sea salt origin of a large proportion of the sulphate in the fog. Comparison of the Cl^-/Na^+ ratio (0.83 in fog, 1.17 in sea water) in the sampled fogs also revealed that much Cl^- had been lost from the fog. Similar methods have been used by Smirnioudi and Siskos (1992), who used ionic ratio calculations to show that all sodium and chloride collected over a year from wet and dry deposition in Athens city centre came from sea salt, whereas nitrates and most sulphates originated from the urban area itself.

Herut et al (1995) used $\delta^{34}S$ data and ionic ratios to identify the sources of sulphur in rainwater in Israel. Na^+/Cl^- ratios show that in most samples (out of a total of 17 rain samples collected at 11 locations) these ions have a clear marine origin. Sulphate exhibits a more complex pattern. Higher $\delta^{34}S$ values were found in samples collected along the Mediterranean coast which had relatively high Cl^-/SO_4^{2-} ratios, whereas lower $\delta^{34}S$ values were found inland, for example, in the Negev desert. $\delta^{34}S$ values ranged from 0.0 to 15.3‰. Conveniently, sea spray sulphate has a uniform isotopic composition, with $\delta^{34}S$ of around 20‰; other sulphates have more variable isotopic signatures. Careful studies of isotope and ionic ratios show that 65% of the samples studied here consist of less than half sea spray sulphate. Thus non-sea spray sulphate from offshore dimethyl sulphide or dust (or perhaps even pollution) is implicated.

Pye and Schiavon (1989) used isotopic signatures of sulphur on buildings to try and interpret the source of the sulphur within efflorescences on Cambridgeshire (eastern England) bridges to establish whether road de-icing salts were responsible for the high sulphate levels found in these bridges. $\delta^{34}S$ values within the bridges ranged from 4.7 to 7.1‰, whereas the road salt had values from 15.2 to 18.9‰, and Pye and Schiavon concluded that road salt was not responsible for the sulphate contamination, which seemed to be caused by air pollution. In a similar study, Klemm and Siedel (1996) used sulphur isotopic signatures to investigate sources of sulphur in efflorescences on historical buildings in Freiburg. They found that various sources with mixing effects were probably involved, as no simple pattern emerged. As well as isotopic and ionic ratios, the sizes of sulphate aerosols can be used as a diagnostic tool. Amoroso and Fassina (1983) report that most sulphate aerosols are 0.1–1 μm in diameter; those over this size are usually from sea spray sulphate and sulphate from wind-blown soil. Submicron sulphate particles are mainly formed by the oxidation of sulphur dioxide. Used together, these methods provide an increasingly powerful way of fingerprinting sources of salts.

ENVIRONMENTAL CONDITIONS FAVOURING SALT WEATHERING

From the discussion of sources of salts and moisture in the preceding sections of this chapter, we can see that a generous supply of salts, coupled with enough moisture to facilitate chemical reactions, is a prerequisite for salt weathering. Thus areas rich in salts from coastal, evaporitic and/or pollution sources are likely to be key sites for the salt weathering hazard. However, as we will see in Chapters 4 and 5, other environmental factors also encourage salt action, such as large and frequent cycling of temperature and humidity (which will facilitate frequent changes of the state of salts), strong evaporation (which will aid the crystallisation of salts and the concentration of salt solutions), high groundwater levels and prevailing winds bringing in salts. These environmental factors are important at a range of scales, with micro-environmental conditions often vitally important at determining whether a particular rock face, wall or desert surface will be prone to salt attack. Increasingly, measurements are being taken of the environmental conditions at sites where the salt hazard is active to aid understanding and prediction. For example, Torfs et al (1996) have carried out environmental monitoring around the church in Sta Marija Ta'Cwerra, Malta. They found that although the church was some distance from the coast, the local topography and high onshore winds produced a high influx of sea salt. Aires-Barros and Mauricio (1996) have used data from the Maltese study, and others within the Mediterranean basin, to develop models to predict the behaviour of salts on monitored walls.

4 Materials' Susceptibility and Experimental Simulations

INTRODUCTION

In Chapter 3 we showed that there is a wide range of salt types present on walls, in lakes and as efflorescences, and many of these types have been implicated in the salt weathering of materials in a range of environmental settings. In this chapter we consider, firstly, the properties of certain materials which make them particularly vulnerable to salt attack, and, secondly, we assemble evidence which will enable us to say something about the relative power of different types, concentrations and conditions of salts. Of particular significance here is the role of experimental simulations in the laboratory, especially those designed to investigate the impacts of salt crystallisation and hydration, the co-association of salt crystallisation and hydration with other weathering processes, especially freeze–thaw, and the operation of sulphation and other linked processes. Field experiments or exposure trials are also worthy of discussion. Such experiments vary hugely in their aims, design and findings, but taken together they act as an invaluable source of information on the weathering roles of salts.

MATERIALS' SUSCEPTIBILITY

Several important physical and chemical properties of rocks and other materials render them susceptible to various types of salt attack, namely the ease with which water can enter, circulate and remain within the material; the fabric, or arrangement of different components of the material; and the geochemical and mineralogical characteristics.

Looking firstly at water penetration, porosity and permeability are highly important parameters relating to the basic rock or material properties, plus any changes wrought by weathering. Porosity may be divided into two main types, i.e. open and closed pores which are, respectively, those with and without access to other pores and the rock or material surface. Pore spaces occur at a wide range of scales, from minute crevices within crystal lattices, through pores within and between particles or grains within the material. Characteristically, porosity is defined as the whole open space available in a rock with a maximum size of a few millimetres (Fitzner, 1994) and thus does not include large fissures, cracks and openings which may, nevertheless, be important in the course of salt weathering. Components of porosity commonly studied include: the total porosity volume; the total pore surface area; the pore shape; and the pore size distribution. A range of methods is available for such investigations, including optical microscopy of thin sections (used for pores of about 0.005–1 mm diameter), scanning electron microscopy (SEM; used for pores from about 0.1 μm–1 mm in diameter); mercury porosimetry (used for pores of about 0.005 μm–0.5 mm) and nitrogen adsorption (for pores of about 0.001–0.1 μm).

The porosity characteristics of a material affect salt weathering in two main ways. They affect how water moves in and out of the material through inflow, capillarity and evaporation and they affect where the major foci of salt crystallisation are. Salts crystallise in larger pores first for thermodynamic reasons, only crystallising in small pores once the large ones are filled. Stones with a high total porosity, with large pores and many small pores, are particularly susceptible to salt weathering.

Several workers have shown how the porosity characteristics of a material relate directly to its susceptibility to salt weathering attack. Laue et al (1996), for example, studied the porosity of three different building stones suffering from extreme salt deterioration in the crypt of St Maria im Kapitol, Cologne, Germany (sandstone, bioclastic limestone and trachyte). The sandstone is decaying by granular disintegration; the bioclastic limestone has 1–3 mm thick salt crusts on the surface, which sometimes detach, pulling off fragments of the limestone as well; and the trachyte exhibits a cracked and scaling surface. In the sandstone, which has a relatively high content of macropores interconnecting microporous patches in the cement, salt crystallisation in the small pores is encouraged, leading to the detaching of individual grains. In the limestone, which has a very fine, well-connected intergranular pore system, the crystallisation of salts occurs mainly on the surface. By contrast, the trachyte has a heterogeneous pore structure, characterised by small fissures near weathered phenocrysts, which encourages the scaling of fragments.

Kozlowski et al (1990) investigated the role of porosity and other features of the physical make-up of stone in explaining the weathering behaviour of Pinczów limestone in Cracow, where adjoining blocks are often found to have very different degrees of damage. Two main types of Pinczów limestone were identified: a coarse-grained, well-preserved variety, and a fine-grained, decayed variety. Their porosity was studied by mercury porosimetry and optical microscopy. Both types are characterised by macropores (i.e. pores >0.05 μm), but the fine-grained variety has a higher total porosity (about 31%) than the coarse-grained variety (about 17%). Optical microscopy revealed that the coarse-grained variety has a porosity dominated by large pores (in the mm range), whereas the fine-grained variety has smaller pores, often with narrow necks (thereby aiding water retention in the pores). Kozlowski et al (1990) state that the open pore structure characteristic of the coarse-grained limestone encourages the easy ingress and removal of water and soluble salts, thereby limiting the damage done. In contrast, water filling the pores of the fine-grained limestone is retained easily, and evaporation near the surface increases the concentration of the remaining fluids, thereby increasing their damaging effect.

Fabric exerts a linked control over salt weathering damage to that of porosity. Fabric includes such features of a material as grain size, cement types and larger scale features such as bedding planes and other structures. Some of these components of fabric provide planes of weakness within the material, affecting its strength (and thus its resistance to salt weathering); others influence the effective surface area exposed to salt attack). Kozlowski et al's (1990) study also considered fabric as an influence on the decay of Pinczów limestone and found that, within the fine-grained variety, those samples that showed considerable recrystallisation of micrite cement into sparite (larger sized calcite crystals) were less decayed than those dominated by micrite.

The geochemical make-up of materials also exerts an influence on their susceptibility to salt attack. As we have seen in Chapter 3, rocks containing calcium carbonate are particularly vulnerable to sulphation where sulphur (from air pollution and other sources) reacts with the calcium carbonate to form gypsum. Siliceous rocks are not prone to such attack in general. Similarly, several studies have shown that granite may be susceptible to salt attack in conjunction with other weathering processes when micas and feldspars become weathered, increasing the porosity of the rock. Thus, Jones et al (1996) found, in a detailed study of weathered Leinster granite from Trinity College, Dublin, that gypsum was present in cracks and cleavage planes within muscovite mica. They inferred that the gypsum, through crystallisation and dissolution cycling, was at least partly responsible for the formation and widening of such cracks. Schiavon (1993), studying decayed granites from monuments in Spain and Portugal, found a different situation. Here, although gypsum crystallisation within samples was common, bio-deterioration seemed to be implicated in the initial increase in intra- and inter-crystalline porosity, which then allows the ingress of salt solutions and the precipitation of damaging gypsum. Together, these two processes seem to be responsible for producing a network of microfractures in the stone, contributing to the detachment of granitic scales.

EXPERIMENTAL SIMULATIONS OF
SALT CRYSTALLISATION AND HYDRATION

Before discussing the results of some of these simulations it is important to note that it is extremely difficult to draw precise conclusions or to make meaningful comparisons from the different studies because the methods used, the materials tested, the environments simulated and the techniques for describing breakdown have been highly variable. It is only relatively recently that sophisticated environmental cabinets have been used which allow the

accurate control of temperature and humidity cycles and which enable these
cycles to give a reasonable replication of what takes place in the real world.
Equally, rock properties have often been described in only the most scant of
ways, and sample dimensions have often not been standardised (see Goudie,
1974 and Robinson and Williams, 1982 for a discussion). Again it is only
relatively recently that techniques such as dilation measurement, ultrasonic
testing and the use of the scanning electron microscope have enabled a precise
discussion of the response of rock pores to salt attack (e.g. Weiss, 1992; Prick,
1996). Single salts, rather than salts in combination, and saturated solutions,
rather than more dilute treatments, have been the norm, while (as Table 4.1
shows) most simulations have involved cyclic or continuous immersion rather

Table 4.1. Some previous salt weathering simulation methodologies

Reference	Method of measuring change	Treatment
Sperling and Cooke (1985)	Loss of weight	(i) Continuous immersion in saturated solutions
		(ii) One initial immersion cycle
Pye and Sperling (1983)	Grain size change	Continuous immersion in saturated solutions
Kwaad (1970)	Weight loss and grain size change	Continuous partial immersion in saturated solutions
Cooke (1979)	Weight loss and degree of splitting	Cyclic immersion in saturated solutions
Jerwood et al (1990b)	Weight loss and particle size change	One immersion cycle with frost Miscellaneous salt concentrations
Jerwood et al (1990a)	Weight loss and particle size change	Continuous immersion in miscellaneous salt concentrations
Smith and McGreevy (1983)	Visual appraisal of surface disagregation	One initial immersion cycle in miscellaneous salt concentrations
Fahey (1985)	Changes in particle size	Low salt concentrations (0.2 M). Partial immersion
McGreevy (1982)	Weight loss	Cyclic immersion in miscellaneous salt concentrations (including saturated)
Goudie (1986)	Weight loss and particle size change	Continuous immersion with salts of miscellaneous concentrations
Goudie et al (1970)	Weight loss and splitting	Cyclic immersion in saturated solutions
Goudie (1974)	Weight loss	Cyclic immersion in saturated solutions
Goudie (1993)	Debris liberation	Single immersion followed by atmospheric cycles

than non-immersion techniques which are probably more representative of the situation on most walls or rock outcrops.

Tests for building stone variability or aggregate soundness (Minty, 1965; Minty and Monk, 1966), often called "crystallisation tests", tend to be particularly *ad hoc* in their conception and have many of the limitations alluded to above. This is clear when we consider the crystallisation test used over very many years by Britain's Building Research Establishment (Ross and Butlin, 1989: 3), the steps in which are described below.

1. A stock solution of sodium sulphate is made up by dissolving 1.4 kg of $Na_2SO_4 \cdot 10H_2O$ in 8.6 l of water. The temperature needs to be kept at $20 \pm 0.5°C$ throughout the test.
2. Saw rocks into 4 cm cubes for testing. At least four samples of each stone should be used for the tests, but unless the stone is particularly variable a maximum of six should be sufficient.
3. Wash with fresh water to remove any loose material, and dry the samples to constant weight at $103 \pm 2°C$.
4. Remove samples from the oven, allow them to cool to $20 \pm 2°C$ and then weigh to ± 0.01 g (W_0).
5. Label the samples and weigh them again (W_1).
6. Place each sample in a 250-ml container, cover with fresh sodium sulphate solution to about 8 mm depth and leave for 2 h at a constant room temperature of $20 \pm 0.5°C$. After the samples have soaked for 1.5 h, place a shallow tray containing 300 ml of water in the oven.
7. After a total of 2 h of soaking the samples should be removed from the solution, placed in the oven on wire racks, and dried for 16 h at $103 \pm 2°C$.
8. Remove samples from the oven and allow them to cool to $20 \pm 2°C$. Note that steps 6–8 constitute one cycle of the test and should be repeated until 15 cycles have been completed.
9. Weigh the samples (W_f) and calculate the percentage weight loss for each sample using the expression
$$\% \text{ weight loss} = 100(W_f - W_1/W_0)$$
10. Calculate the mean percentage weight loss for each set of samples.

We might ask questions about the basis for the concentration of sodium sulphate used, why sodium sulphate was used, why the humidity is not carefully controlled, why such an extreme drying temperature is used, and why the soaking phase has the length it has. Similar questions might be asked of the standard German (DIN 52 111 and VDI 3797) and American (AM C88-76) durability tests.

Nonetheless, for all their imperfections, laboratory simulations have enabled us to make some major strides in understanding the role of different salts and the response of different rock types.

During the 1950s some important work was undertaken in France to simulate salt weathering in the laboratory, primarily using granites and lavas. The work demonstrated that granite was susceptible to salt attack (Birot, 1954). It also provided some of the first evidence for the relative powers of different types of salt. Pedro (1957a) subjected granite blocks to a range of solutions and dried them at 80°C over a four month period. The percentage of disintegration achieved by different salts was as follows: $NaNO_3$, 0.79; Na_2SO_4, 0.67; $Mg(NO_3)_2$, 0.66; K_2SO_4, 0.40; KNO_3, 0.33; Na_2CO_3, 0.28; K_2CO_3, 0.20; $MgSO_4$, 0.20; $CaSO_4$, 0.14; $Ca(NO_3)_2$, 0.04; and H_2O, 0.04.

Another important simulation was undertaken by Kwaad (1970) in the Netherlands. He partially immersed granite samples and subjected them to a range of temperature and humidity cycles in a climatic cabinet. His main experiment alternated every 12 hours between a temperature of 15°C with a relative humidity of 90% and a temperature of 70°C with a relative humidity of 10%. Under these extreme cyclic conditions, saturated solutions of different salts were used and the ranking of effectiveness from most to least was: sodium sulphate; sodium carbonate; magnesium sulphate; sodium chloride; and calcium sulphate. Sodium sulphate was the most effective by a long way, while calcium sulphate had no perceptible effect.

In the same year Goudie et al (1970) performed some primitive experiments. Rock samples of approximately similar initial weight, shape and size were immersed in saturated salt solutions at 17°C for one hour. They were then removed and dried in an oven, using a temperature of 60°C for six hours and a temperature of 30°C for the remainder of the 24 hour cycle. The procedure was repeated 40 times and no attempt was made to control the humidity. The rock used was Arden sandstone. The ranking of effectiveness from most to least was: sodium sulphate; magnesium sulphate; calcium chloride; sodium carbonate; and sodium chloride.

This work was extended by Goudie (1974). He used 3 cm cubes of chalk and York stone which were immersed in various saturated salt solutions at 17–20°C for one hour. They were then removed and dried in an oven, using a temperature of 60°C for six hours and of 30°C for the remainder of the 24 hour cycle. For the York stone, 58 cycles were used, whereas for the chalk 43 cycles were used.

Figures 4.1 and 4.2 show the speed with which the two rocks, sandstone and chalk, broke down when subjected to the various process treatments outlined. Tables 4.2 and 4.3 exemplify this further by ranking the processes in descending order of effectiveness.

The most striking feature of both the tables and the figures is the fact that salt crystallisation (Na_2SO_4, $MgSO_4$, $CaCl_2$ and Na_2CO_3 for sandstone; Na_2SO_4, Na_2CO_3, $NaNO_3$ and $CaCl_2$ for chalk) appears clearly as the most effective of the treatments utilised. By contrast, frost, though causing some breakdown, was relatively ineffective. Insolation, thermal expansion of salt, and wetting and drying caused no measurable change in sample weight over the cycles utilised (58 cycles for sandstone, 43 for chalk). The ineffectiveness

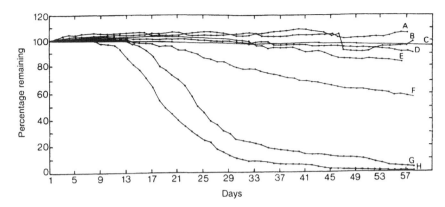

Figure 4.1. The progress of breakdown of silica-cemented sandstone under the action of different treatments. A = Sodium nitrate crystallisation; B = sea salt crystallisation; C = frost; D = sodium carbonate crystallisation; E = sodium chloride crystallisation; F = magnesium sulphate crystallisation; G = sodium sulphate with daily leaching; and H = sodium sulphate crystallisation. From Goudie (1974: figure 1). Reproduced with permission

of the thermal expansion of salt was confirmed by the 2160 cycle treatment in which, once again, there was no detectable loss in weight or development of cracks.

Certain salts, notably sea brine and $NaNO_3$ for sandstone, and sea brine and $NaCl$ for chalk, proved to be ineffective; indeed, with the treatment described so far, samples gained in weight on account of the development of a salt efflorescence of considerable volume. It was, however, noted that the blocks thus treated, although increased in weight, did develop blisters and crack. Thus at the conclusion of the cycles these samples were placed in hot water for 48 hours to leach out the salt, which in some cases appeared to be binding fractured blocks together. This resulted in the total disintegration of certain samples. Using these results another ranking of samples in terms of effectiveness of different salts can be obtained (Tables 4.4 and 4.5), which serves to underline the effectiveness of salt crystallisation compared with other processes.

Smith and McGreevy (1983) reported a simple experiment in which 10% solutions of $MgSO_4$, $NaCl$ and Na_2SO_4 were used to break down sandstone samples during a diurnal regime of heating and cooling of between 53 and 21°C. For 40 cycles, the amount of debris liberated in g cm^{-2} was as follows for bedding normal to exposure: $MgSO_4$, 0.020; Na_2SO_4, 0.071; and $NaCl$, 0.010. For bedding parallel to exposure the amount of debris was: $MgSO_4$, 0.012; Na_2SO_4, 0.012; and $NaCl$, 0.004. Once again sodium sulphate is the most effective salt, while sodium chloride achieves relatively little.

One of the most comprehensive attempts to assess the relative efficiencies of different salts, different concentrations of salts and different combinations

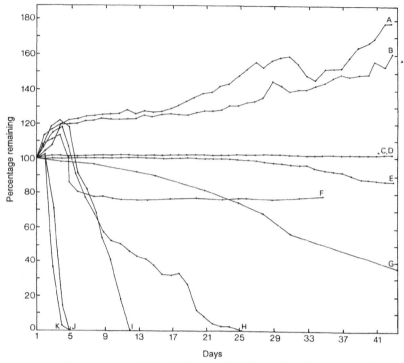

Figure 4.2. The progress of breakdown of chalk under the action of different treatments. A = Sea salt crystallisation; B = sodium chloride crystallisation; C and D = insolation and wetting and drying; E = calcium sulphate crystallisation; F = magnesium sulphate crystallisation; G = frost; H = calcium chloride crystallisation; I = sodium nitrate crystallisation; J = sodium carbonate crystallisation; and K = sodium sulphate crystallisation. From Goudie (1974: figure 2). Reproduced with permission

Table 4.2. Relative resistance of silica-cemented sandstone to various mechanical weathering processes

Ranking	Treatment	Loss after 58 cycles (%)
1	Salt crystallisation (Na_2SO_4)	98.91
2	Salt crystallisation (Na_2SO_4 with leaching)	95.67
3	Salt crystallisation ($MgSO_4$)	35.50
4	Salt crystallisation ($CaCl_2$)	15.84
5	Salt crystallisation (Na_2CO_3)	7.88
6	Frost	2.27
7	Sea brine salt crystallisation	0.33
8=	Insolation	0.00
8=	Wetting and drying	0.00
8=	Thermal expansion of salt ($NaNO_3$)	0.00
11	Sea brine salt crystallisation with leaching	(+1.91)
12	Salt crystallisation ($NaNO_3$)	(+7.57)

Table 4.3. Relative resistance of chalk to various mechanical weathering processes

Ranking	Treatment	Loss after 43 cycles (%)
1	Salt crystallisation (Na_2SO_4)	100 (all gone after 4 cycles)
2	Salt crystallisation (Na_2CO_3)	100 (all gone after 4 cycles)
3	Salt crystallisation ($NaNO_3$)	100 (all gone after 4 cycles)
4	Salt crystallisation ($CaCl_2$)	100 (all gone after 4 cycles)
5	Frost	65.41
6	Salt crystallisation ($MgSO_4$)	23.82
7	Salt crystallisation ($CaSO_4$)	13.19
8=	Insolation	0.00
8=	Wetting and drying	0.00
10	Sea brine crystallisation	Gain
11	NaCl	Gain

of salts was undertaken by Goudie (1986). He used blocks of York stone (a lower Carboniferous siliceous sandstone). These were 180 mm long and were placed in solutions that were topped up each day to maintain a level of 15 mm: 165 mm of rock were therefore available in which salt and moisture migration ("the wick effect") could occur. The temperature cycle consisted of seven hours at 55°C and the rest of the 24 hour cycle at 22°C. No attempt was made to control the humidity. The salt treatments and the breakdown they achieved after 40 and 60 cycles are shown in Table 4.6. After 40 cycles sodium carbonate emerged as the most effective, followed by magnesium sulphate and sodium sulphate. The power of the mixtures of salts in turn reflects the power of these three individual salts, and thus especially effective

Table 4.4. Relative resistance of silica-cemented sandstone to various mechanical weathering processes followed by leaching of remaining salt

Ranking	Treatment	Loss after 58 cycles and then thorough leaching (%)
1	Salt crystallisation (Na_2SO_4)	99.00
2	Salt crystallisation (Na_2SO_4+daily leaching)	95.88
3	Salt crystallisation ($MgSO_4$)	47.19
4	Salt crystallisation ($CaCl_2$)	22.78
5	Salt crystallisation (sea brine)	17.78
6	Salt crystallisation (sea brine with leaching)	14.31
7	Salt crystallisation ($NaCO_3$)	9.93
8	Salt crystallisation ($NaNO_3$)	3.36
9	Frost	2.27
10=	Insolation	0.00
11=	Wetting and drying	0.00
12=	Thermal expansion of salt ($NaNO_3$)	0.00

Table 4.5. Relative resistance of chalk to various mechanical weathering processes followed by leaching of remaining salt

Ranking	Treatment	Loss after 43 cycles with thorough leaching (%)
1	Salt crystallisation (NaSO$_4$)	100 (all gone after 4 cycles)
2	Salt crystallisation (Na$_2$CO$_3$)	100 (all gone after 4 cycles)
3	Salt crystallisation (NaNO$_3$)	100 (all gone after 11 cycles)
4	Salt crystallisation (CaCl$_2$)	100 (all gone after 24 cycles)
5	Salt crystallisation (sea brine)	100 (after 43 cycles and subsequent leaching)
6	Frost	65.41
7	Salt crystallisation (MgSO$_4$)	28.09
8	Salt crystallisation (CaSO$_4$)	14.95
9	Salt crystallisation (NaCl)	10.15
10	Wetting and drying	0.22
11	Insolation	0.00

Source: Goudie (1974: tables 3, 4, 5 and 6).

mixes were those of sodium carbonate and magnesium sulphate, and of magnesium sulphate and sodium sulphate. In general, whether individually or as components of mixes, sodium chloride, sodium nitrate and calcium sulphate were much less potent.

After 60 cycles the ranking was similar to that after 40 cycles, with sodium carbonate, sodium sulphate, magnesium sulphate and their mixes once again being clearly dominant.

Given that it is likely that different salts will have different degrees of effectiveness according to the ambient climate conditions which control hydration and hygroscopicity, Goudie (1993) attempted to assess the role of different climate cycles in determining the relative power of different salts. He used six published cycles of observed ground surface temperature and humidity conditions during a full 24 hour cycle. The material used was concrete, which was subjected to one phase of immersion in saturated solutions of salts before being emplaced into an environmental cabinet which was able to reproduce the six cycles (Figure 4.3).

The Wadi Digla cycle is the same as that used in a previous salt weathering simulation (Cooke, 1979) and is based on observations made by Williams (1923) on 9 August 1922 at Wadi Digla near Cairo, Egypt.

Williams (1923) recorded the rock surface temperature at hourly intervals using a black-bulb thermometer, and also undertook hourly recordings of relative humidity using a wet-and-dry bulb thermometer and a sling psychrometer. The cycles represent extreme warm desert summer conditions, characterised by a great daily march of temperature and low but variable relative humidity. The temperature cycle, though severe, is probably

Table 4.6. Breakdown achieved by different salt treatments

Rock sample	Salt type	Salt concentration $(g^{-1}l)$	Breakdown parameters*					
			1	2	3	4	5	6
A: After 40 cycles:								
2	NaCl	25	100.39	100.39	1	0.21	17.50	25
4	NaCl ·	50	100.53	100.53	1	0.12	17.75	26
6	Na_2CO_3	100	100.47	100.47	1	0.33	16.75	22
8	Na_2CO_3	25	49.99	95.96	2	4.13	5.00	5
10	Na_2CO_3	50	35.58	88.73	5	6.56	2.00	1
12	$NaCO_3$	100	76.36	87.88	3	6.64	4.50	4
14	Na_2SO_4	25	98.09	100	1	2.39	11.75	12
16	Na_2SO_4	50	51.34	95.86	2	3.83	5.50	6
18	Na_2SO_4	00	63.75	97.31	2	3.57	7.00	9
20	$MgSO_4$	25	54.61	94.39	2	2.10	6.25	7=
22	$MgSO_4$	50	51.07	94.35	4	2.73	4.25	3
24	$MgSO_4$	100	55.38	95.37	4	2.08	6.25	7=
26	$NaNO_3$	25	100.64	100.64	1	0.08	18.50	29
28	$NaNO_3$	50	100.08	100.08	1	0.10	18.25	29
30	$NaNO_3$	100	100.26	100.26	1	0.24	17.25	24
32	Sea water	25	100.64	100.64	1	0.29	17.00	23
34	Sea water	50	100.85	100.85	1	0.54	16.50	21
36	Sea water	100	100.70	100.70	1	0.96	15.25	17
39	$NaCl + Na_2CO_3$	50 + 50	99.93	99.93	1	0.58	15.50	18
41	$NaCl + Na_2SO_4$	50 + 50	99.78	99.78	1	0.66	14.25	16
43	$NaCl + NaNO_3$	50 + 50	100.00	100.00	1	0.03	18.75	30
45	$Na_2CO_3 + Na_2SO_4$	50 + 50	99.75	99.75	1	0.72	13.50	14
47	$Na_2CO_3 + NaNO_3$	50 + 50	99.90	99.90	1	10.00	14.00	15
49	$Na_2SO_4 + NaNO_3$	50 + 50	100.00	100.00	1	0.55	16.25	20
51	$NaCl + MgSO_4$	50 + 50	98.89	98.89	1	1.85	11.25	11
53	$Na_2CO_3 + MgSO_4$	50 + 50	41.05	76.36	2	15.40	2.25	2
55	$Na_2SO_4 + MgSO_4$	50 + 50	99.72	99.72	1	1.51	12.25	13
57	$NaNO_3 + MgSO_4$	50 + 50	64.03	98.21	2	1.88	9.00	10
59	$NaCl + CaSO_4$	225 + 10.24	100.00	100.00	1	0.56	16.00	19
61	NaCl	225	100.00	100.00	1	0.12	17.75	27
62	$CaSO_4$	2	100.00	100.00	1	0.00	19.00	31=
37	Deionised H_2O	0	100.00	100.00	1	0.00	19.00	31=
B: After 60 cycles								
1	NaCl	25	99.83	99.83	1	0.66	18.50	19
3	NaCl	50	100.02	100.02	1	0.47	20.50	22=
5	NaCl	100	100.13	100.13	1	0.53	32.25	31
7	Na_2CO_3	25	50.20	93.93	2	6.96	7.00	6
9	Na_2CO_3	50	46.91	81.18	3	9.85	3.55	3
11	Na_2CO_3	100	34.22	82.77	3	11.36	1.75	1
13	Na_2SO_4	25	49.06	90.33	2	9.41	7.25	7
15	Na_2SO_4	50	44.21	91.00	3	7.76	4.50	4

Continued

Table 4.6. *continued*

Rock sample	Salt type	Salt concentration $(g^{-1}l)$	Breakdown parameters*					
			1	2	3	4	5	6
17	Na_2SO_4	100	62.73	94.68	2	4.90	10.00	11=
19	$MgSO_4$	25	96.65	96.65	1	3.75	10.00	11=
21	$MgSO_4$	50	52.08	95.40	3	3.53	8.50	9
23	$MgSO_4$	100	59.25	85.51	2	6.33	7.75	8
25	$NaNO_3$	25	100.55	100.55	1	0.20	25.25	28=
27	$NaNO_3$	50	100.54	100.54	1	0.13	25.25	28=
29	$NaNO_3$	100	100.62	100.62	1	0.46	25.25	28=
31	Sea water	25	99.92	99.92	1	0.81	18.75	20=
33	Sea water	50	100.10	100.10	1	0.98	20.50	22=
35	Sea water	100	100.32	100.32	1	1.24	21.00	24
38	$NaCl + Na_2CO_3$	50 + 50	44.42	81.58	4	4.99	5.25	4.75
40	$NaCl + NaSO_4$	50 + 50	55.21	96.57	2	2.15	11.00	13
42	$NaCl + Na_2CO_3$	50 + 50	100.43	100.43	1	0.19	24.50	27
44	$Na_2CO_3 + Na_2SO_4$	50 + 50	98.18	98.18	1	2.28	14.00	15
46	$Na_2CO_3 + NaNO_3$	50 + 50	56.41	98.49	2	0.88	13.25	14
48	$Na_2SO_4 + NaNO_3$	50 + 50	100.02	100.02	1	1.13	18.75	20=
50	$NaCl + MgSO_4$	50 + 50	98.94	98.94	1	2.13	15.75	17
52	$Na_2CO_3 + MgSO_4$	50 + 50	38.21	74.20	3	16.63	2.00	2
54	$NaNO_3 + MgSO_4$	50 + 50	98.87	98.87	1	2.54	14.50	16
56	$Na_2SO_4 + MgSO_4$	50 + 50	62.10	96.20	2	5.00	9.75	10
58	$NaCl + CaSO_4$	225 + 10.24	99.70	99.70	1	0.94	17.25	18
60	NaCl	225	100.03	100.03	1	0.38	22.00	26
62	$CaSO_4$	2	99.98	99.98	1	0.08	21.5	25

*1, Weight of largest remaining particle as percentage of original weight; 2, control weight of all particles >2 g as percentage of original weight; 3, number of particles produced with weight >2 g; 4, percentage of material produced with diameter of <1 mm; 5, index of disintegration; and 6, overall ranking.

reasonably typical of that found in many warm desert environments in the summer months. The daytime peak of 72°C is, for example, comparable with measured rock surface temperature values for Tibesti (Peel, 1974), Sudan (Cloudsley-Thompson and Chadwick, 1964), Arizona (Sinclair, 1922) and California (Chappell and Bartholomew, 1981).

The Negev cycle is for 26 and 27 September 1967 at Avdat, in the Negev Desert of Israel (Orshan, 1986). The daily temperature range is between 10 and 41°C and the humidity range is between 20 and 100%, with the highest humidities coinciding with the low nocturnal temperatures. This cycle involves, in effect, the development of a heavy dew, a phenomenon known to be of frequent occurrence in the Negev Desert.

The Gobabeb cycle is for 4 July 1976 at Gobabeb in the Namib Desert, Namibia. The daily temperature change is similar to that for the Negev cycle,

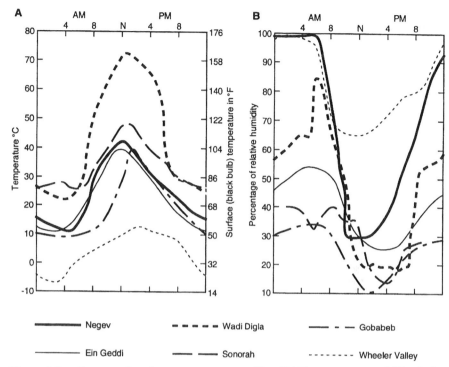

Figure 4.3. Six ground surface temperature profiles. (A) Temperature; and (B) relative humidity. Details of sources are given in the text (from Goudie, 1993: figure 1). Reproduced with permission

being between 9 and 39°C. However, the relative humidity change is much less, ranging between 11% in the early afternoon and 34% in the early morning. The data were provided by Dr M. Seely (personal communication).

The Ein Geddi cycle is for 19 January 1961 at Ein Geddi in Israel (Warburg, 1964). The temperature range is comparable with that of the Negev and Gobabeb cycles, being between 11 and 38°C. However, the relative humidity changes are intermediate in character, ranging from 26% in the late afternoon to 53% in the early morning.

The Sonorah cycle is from Sonorah in Mexico for 23 June 1971 (Friedman, 1980). The temperature range is intermediate between that of the Wadi Digla cycles on the one hand, and those of Negev, Gobabeb and Ein Geddi on the other. It ranges between 26 and 48°C. The relative humidity ranges between 18% in the early afternoon to 40% during the night.

The sixth cycle provides a contrast to the other five, being for the Wheeler Dry Valley in Antarctica for 29 December 1967 (Cameron, 1969). This is a very cold dry desert. This mid-summer cycle varies from −7 to +13°C. The nocturnal relative humidity reaches a very high value (99%) and falls to 66% at noon.

Table 4.7. Salt solubilities and salt uptake

Salt	Solubility at 25°C (gl^{-1})*	Ranking	Average salt uptake (%)[†]	Ranking
Na_2SO_4	219.0	5	5.56	5
$MgSO_4$	267.0	3	12.18	3
$NaCl$	264.3	4	13.91	2
Na_2CO_3	316.9	2	11.62	4
$NaNO_3$	476.0	1	16.28	1

*Data from Stephen and Stephen (1963).
[†]Difference between weights of rocks before and after salt uptake expressed as a percentage of their initial weight. Mean of 18 samples.
Source: Goudie (1993: table II).

Because of their varying solubilities (Table 4.7) the saturated solutions of different salts will be absorbed to varying degrees when the concrete blocks are immersed in them. The most soluble salt is sodium nitrate and the least soluble is sodium sulphate. Thus the salt uptake was greatest for sodium nitrate and least for sodium sulphate. However, there is not a prefect correlation between solubility and salt uptake, and the uptake of sodium sulphate is less than might be expected from its solubility. The explanation for these differences is not clear, but has been observed by other workers (e.g. McGreevy, 1982).

The results for each cycle type and salt treatment are shown in Table 4.8.

Wadi Digla cycle

This cycle produced the greatest weight loss of any one treatment. The samples treated with sodium nitrate experienced a debris liberation of 3.76%. A significant amount of debris was also liberated by the sodium carbonate treatment (0.78%). The samples treated with sodium chloride, sodium sulphate and magnesium sulphate produced no debris.

Negev cycle

Debris liberation was achieved by four treatments under this cycle. The most effective salt was sodium sulphate, with a debris liberation of 2.15%, the second highest amount for any of the treatments in this experiment. Also effective were sodium carbonate (1.40%), sodium nitrate (0.43%) and magnesium sulphate (0.08%).

Gobabeb cycle

In contrast with the two previous cycles, no debris liberation was achieved by any of the salt treatments.

Table 4.8.   Results of salt experiments: percentage debris liberation after 25 cycles

Salt	Wadi Digla	Negev	Gobabeb	Ein Geddi	Sonorah	Wheeler Valley
Na_2SO_4	0.00	2.71	0.00	0.00	0.00	0.00
	0.00	2.11	0.00	0.00	0.00	0.00
	0.00	1.62	0.00	0.00	0.00	0.00
Mean	0.00	2.15	0.00	0.00	0.00	0.00
$MgSO_4$	0.00	0.16	0.00	0.00	0.00	0.00
	0.00	0.01	0.00	0.00	0.00	0.00
	0.00	0.08	0.00	0.00	0.00	0.00
Mean	0.00	0.08	0.00	0.00	0.00	0.00
NaCl	0.05	0.00	0.00	0.00	0.00	0.00
	0.06	0.00	0.00	0.00	0.00	0.00
	0.06	0.00	0.00	0.00	0.00	0.00
Mean	0.06	0.00	0.00	0.00	0.00	0.00
Na_2CO_3	0.15	1.60	0.00	0.00	0.00	0.24
	0.49	1.35	0.00	0.00	0.00	0.21
	0.70	1.24	0.00	0.00	0.00	0.23
Mean	0.78	1.40	0.00	0.00	0.00	0.22
$NaNO_3$	3.34	0.20	0.00	0.00	0.14	0.00
	4.24	0.88	0.00	0.00	0.14	0.00
	3.71	0.22	0.00	0.00	0.00	0.00
Mean	3.76	0.43	0.00	0.00	0.14	0.00

Source: Goudie (1993: table III).

Ein Geddi cycle

As with the Gobabeb cycle, no debris liberation was achieved by any of the salt treatments.

Sonorah cycle

A small amount of debris liberation (0.14%) was achieved by the sodium nitrate treatment. No other treatment was effective.

Wheeler Valley cycle

A very small amount of debris liberation (0.22%) was achieved by the sodium carbonate treatment. No other treatment was effective.

Discussion of experiments

What emerges from this experiment is that different salts are indeed effective under different climate conditions. The most effective salt under Wadi Digla conditions was sodium nitrate. This can be explained in terms of the threshold for the transition from deliquescence to crystallisation for the salt taking place only at a relatively high temperature (42°C) (Pantony, 1961). It is perhaps significant that sodium nitrate also caused some debris liberation with the Sonorah cycle, which reaches 48°C.

By contrast, under Negev conditions, sodium sulphate was effective, in spite of the fact that it was the salt least taken up into samples. The effectiveness of the Negev conditions may be explained by the high atmospheric humidity levels that occur at times of low temperature. This allows the hydration of hydratable salts. For example, for thenardite (Na_2SO_4) to hydrate to mirabilite ($Na_2SO_4 \cdot 10H_2O$) at 20°C, the relative humidity must exceed 71%, while for the hydrates of magnesium sulphate the critical relative humidity value lies between 80 and 90% (Fahey, 1985). The Negev cycle is the only one of the six where these conditions are met. Both sodium carbonate and sodium sulphate show particularly high volume expansions on hydration and they were the two most effective salts under Negev cycle conditions.

The only salt that was effective under Wheeler Valley cycle conditions was sodium carbonate. It achieved some debris liberation when other salts did not. A possible reason for this is that the hydration pressures that are set up by this salt are particularly high at low temperatures and high relative humidities (Winkler and Wilhelm, 1970).

Only one salt, sodium chloride, had no effect under any of the six cycles used. It is the only salt of those used that either does not hydrate appreciably over the range of environments simulated (sodium sulphate, magnesium sulphate, sodium carbonate) or have the temperature-related deliquescence characteristics of sodium nitrate.

If we consider all these experiments together (Table 4.9), we can see that there is some variability in ranking. However, certain general points emerge.

Table 4.9. Experimental ranking of the power of different salts (most effective salt at top of column)

Pedro (1957a)	Kwaad (1970)	Goudie et al (1970)	Goudie (1974)	Goudie (1974)	Goudie (1986)	Goudie (1993) (Wadi Digla Cycle)	Goudie (1993) (Negev Cycle)
$NaNO_3$	Na_2SO_4	Na_2SO_4	Na_2SO_4	Na_2SO_4	Na_2CO_3	$NaNO_3$	Na_2SO_4
Na_2SO_4	Na_2CO_3	$MgSO_4$	$MgSO_4$	Na_2CO_3	$MgSO_4$	Na_2CO_3	Na_2CO_3
$Mg(NO_3)_2$	$MgSO_4$	$CaCl_2$	$CaCl_2$	$NaNO_3$	Na_2SO_4		$NaNO_3$
K_2SO_4	$NaCl$	Na_2CO_3	Na_2CO_3	$CaCl_2$	$NaCl$		$MgSO_4$
KNO_3	$CaSO_4$	$NaCl$	$NaNO_3$	$MgSO_4$	$NaNO_3$		$NaCl$
Na_2CO_3				$NaCl$	$CaSO_4$		
K_2CO_3							
$MgSO_4$							
$CaSO_4$							
$Ca(NO_3)_2$							

The first of these is that certain salts, most notably sodium sulphate, sodium carbonate and magnesium sulphate, appear to be consistently effective, with sodium sulphate being the most consistent of all. The second is that sodium nitrate can sometimes be effective and sometimes not. The third is that two of the most common salts in the environment, calcium sulphate and sodium chloride appear to be relatively ineffective. The consistently effective salts are those that both hydrate at the temperatures and humidities used in the cycle and which are highly soluble. In addition, sodium sulphate has been identified as possessing particular characteristics that may account for its pre-eminent position (Goudie, 1977).

1. Sodium sulphate undergoes a high degree of volume change from its dehydrated state (Na_2SO_4 — thenardite — density 2.68) to its hydrated state ($Na_2SO_4 \cdot 10H_2O$ — mirabilite — density 1.46). At 1 atm pressure, in the presence of a saturated solution, the equilibrium temperature for the reaction $Na_2SO_4 \cdot 10H_2O = Na_2SO_4 + 10H_2O$ is 32.4°C. The transition temperature decreases significantly in the presence of sodium chloride (as low as 17.9°C in an NaCl-saturated environment) and to a lesser extent with sodium carbonate (Eugster and Smith, 1965). Normal warm desert temperatures and humidity cycles could promote the volume increase (which amounts to 315%) daily. For comparison, the volume increases on hydration for other common salts are listed in Table 5.5. Moreover, as Winkler (1975: 125) writes: "The hydration of the sodium sulphates thenardite to mirabilite is more rapid than hydration of other salts; their hydration and dehydration may repeat several times in a single day: even low hydration pressures may become effective in such rapid change". The dehydration of mirabilite to thenardite does not take longer than 20 minutes at 39°C (see Boulanger and Urbain, 1912 and Mortensen, 1933).

2. The rapid decrease in the solubility of sodium sulphate as the temperature falls from 32.3°C is important. Such a crystallisation of a salt solution on a temperature fall affects a much larger volume of salt per unit time than crystallisation induced by evaporation, which is a gradual process (Kwaad, 1970: 79). The solubility of sodium sulphate is more sensitive to temperature than is that of other common salts such as sodium nitrate, magnesium sulphate, sodium carbonate, sodium chloride and calcium sulphate.

3. Nevertheless, because sodium sulphate is so highly soluble substantial quantities of sulphate are available for the process of crystal growth when solutions are evaporated by high diurnal temperatures. Evaporation would also help to create a saturated solution from which crystals could grow on cooling.

4. This effect of a temperature rise could be compounded by the fact that unlike many compounds, the solubility of sodium sulphate does not increase in a linear fashion with temperature. Above a temperature of about 32°C the solubility is reduced. Thus the strong diurnal heating of

saline solutions could lead to crystal growth. The summer month
temperature range is sufficient in many deserts for this process to operate.
On raising the temperature of a solution containing 100 g of water and
78.6 g of sodium sulphate to 60°C, the anhydrous sodium sulphate would
precipitate out of solution until only 46 g were present in the 100 g of
water (Rueffel, 1968).

5. The needle-shaped nature of the sodium sulphate crystals might tend to
   increase their disruptive power. The crystals of mirabilite are of a very
   long prismatic type (Wells, 1923).

The highly variable behaviour of sodium nitrate in the different experiments
can probably be accounted for in terms of its hygroscopic characteristics as
discussed in relation to Goudie's 1993 experiment. It is a highly deliquescent
salt (as indeed is calcium chloride) and effective crystallisation can only take
place at relatively high ambient temperatures. If temperatures are low it
cannot crystallise and given that it is also a salt that does not hydrate, it is
therefore ineffective.

That brings us to a consideration of the relatively ineffective nature of
calcium sulphate and sodium chloride in the experiments that have been
conducted to date. The prime reason why calcium sulphate appears ineffective
is that it is so sparingly soluble compared with other common salts (Figure
4.4). The reasons why sodium chloride may appear less effective include the
fact that it does not hydrate (except near the freezing point of water), its
solubility is not very temperature dependent (Figure 4.4), it often appears to
cement rather than disrupt rock until it is leached (Goudie, 1974), and because
of its deliquescent characteristics it may not crystallise under certain ambient
conditions.

However, there is abundant field evidence that gypsum and sodium chloride
are *not* ineffective and further experimental simulations need to be developed
to address this issue.

The action of salts in combination may in some cases be greater than the
action of the individual salts in isolation. This has often been suspected in the
widespread case of where gypsum and sodium chloride occur together. One
explanation for this has been provided by Price and Brimblecombe (1994: 92)
in the context of the equilibrium relative humidities (RH) of different salts;
using Figure 4.5, they remark

Whilst sparingly soluble calcium sulphate has virtually no effect on the solubility
of sodium chloride, sodium chloride has a marked effect on the solubility of
calcium sulphate ($CaSO_4 \cdot 2H_2O$). As the concentration of sodium chloride
increases, the solubility of calcium sulphate increases by a factor of four, but then
falls again as the sodium chloride concentration increases further. What is more,
the relative humidity of air in equilibrium with the system covers the range 75%
to 99.86% (99.96% is the equilibrium RH of a saturated calcium sulphate
solution). Not only does sodium chloride increase the solubility of calcium
sulphate, it also causes it to crystallize out as the ambient RH falls between 90%

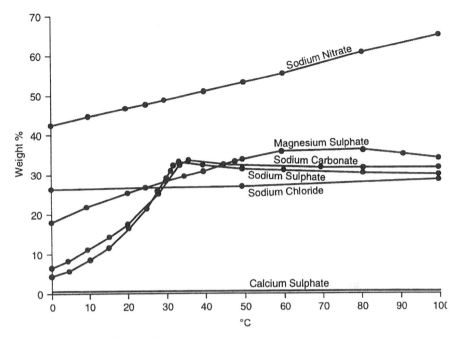

Figure 4.4. Solubility of different salts in relation to solution temperature. From Goudie (1977: figure 6). Reproduced with permission

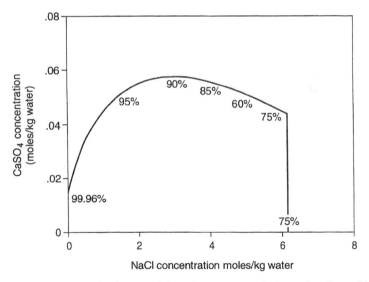

Figure 4.5. Solubility of calcium sulphate in aqueous solutions of sodium chloride at 20°C and vice versa. The numbers beside the lines give the relative humidity of air in equilibrium with the system at those points. Modified after Price and Brimblecombe (1994: figure 6) and reproduced with permission

and 75%, or as it rises above 90%! Whereas calcium sulphate on its own can only cause damage following wetting by rain or mist, calcium sulphate in association with sodium chloride can cause damage whenever the ambient relative humidity fluctuates within the range 75 to 100% — common enough conditions in temperate climates.

There are various other reasons why experimental simulations using a solution of pure gypsum may not replicate conditions in nature, where solutions will tend to have a complex mineralogical composition. The presence of other evaporite minerals can influence the behaviour of gypsum and can accelerate the salt weathering process. Livingston (1994) has discussed the mechanisms by which this can happen. First of all, there is "the deliquescence effect". A pure gypsum crust rarely deliquesces and so, once formed, tends to be relatively inactive. However, when a more deliquescent salt such as sodium chloride is present, the stone surface will become moistened more frequently and thus provide a solution for dissolving, recrystallising and translocating gypsum more often. Secondly, in addition to increasing the frequency of the dissolution/recrystallisation of crystals of gypsum, the presence of additional evaporite minerals can also increase the intensity of each cycle. Livingston (1994: 103) argues that this occurs "because the additional solutes increase the ionic strength of the solution, thereby reducing the activity coefficients of Ca^{2+} and SO_4^{2-}. The lower activities in turn increase the solubility of gypsum and hence the volume of material that is dissolved and then re-precipitated. For example, the solubility of gypsum in a 2 molar solution of NaCl is roughly 5 times that in a pure gypsum solution". Thirdly, the presence of other salts can affect the hydration state of the gypsum, lowering the transitional temperature between the hydrated and non-hydrated form of calcium sulphate. For example, in a 20% sodium chloride solution, the transition temperature may be as low as 25°C (rather than around 58°C for pure gypsum). This greatly increases the likelihood of phase changes occurring.

Finally, whatever salts are used in laboratory simulations of salt weathering, the concentrations of salt in solution are an important control of the rate of rock breakdown and debris liberation. Using the Negev cycle and two different rock types (Hollington stone and Bath stone top bed), the amount of debris liberated after 25 cycles increased exponentially as the concentrations of sodium carbonate and sodium sulphate increased (Figure 4.6). However, such a clear relationship between the amount of debris liberation and salt concentration may not exist when subfreezing conditions are considered.

RELATIVE RESISTANCE OF DIFFERENT ROCKS

In addition to looking at the relative power of different salts, experimental simulations have also been used to look at the resistance of different rocks and

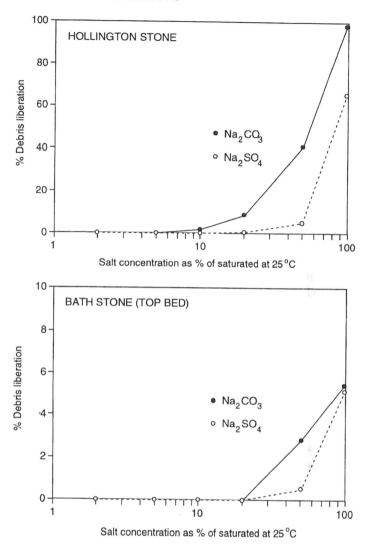

Figure 4.6. Amount of debris liberated from Hollington Stone (sandstone) and Bath Stone Top Bed (a limestone) at various concentrations of solutions of sodium carbonate and sodium sulphate

other building materials to salt attack. Some of the first experiments of this type were those undertaken by Luquer (1895) using sodium sulphate (Table 4.10) on a wide range of rock types. The susceptibility of his sandstone compared with granites, gneisses, marbles and limestones is clear.

Goudie et al (1970) also subjected an array of different rocks to sodium sulphate attack and found (Figure 4.7) that chalk, limestone and sandstones

Table 4.10. Luquer's results relating to resistance of different rock types

Specimens tested	Original weight (g)	Loss of weight (g)	Loss of weight (parts in 10 000)
Coarse crystalline dolomite marble	71.9020	0.0775	10.78
Medium crystalline dolomite marble	93.8861	0.1597	17.01
Fine-grained limestone	67.0964	0.1744	25.99
Coarse-grained granite	71.8648	0.1115	15.51
Medium-grained red granite	56.4939	0.0370	6.55
Fine-grained grey granite	43.5910	0.0225	5.16
Rather fine-grained gneiss	61.8687	0.0392	6.33
Norite "Au Sable" granite	35.1173	0.0135	3.84
Decomposed sandstone	39.4294	1.9010	482.12
Very fine-grained sandstone	37.7760	0.1800	47.65
Sandstone	28.0325	0.4070	145.18
Pressed brick	37.4025	0.0930	24.86
Decomposed sandstone	22.9660	3.7235	1621.31
Sandstone	23.9001	0.1381	57.78

Source: Luquer (1895: 243). Reproduced with permission.

were markedly less resistant than diorite, dolerite and granite. They found that there was a highly significant correlation ($S = 0.9$) between amount of breakdown and the water absorption capacities of the different rocks. This work was extended by Goudie (1974) for a wider range of materials (Table 4.11), and once again there was a highly significant correlation ($S = 0.83$) between the water absorption capacity and the loss of weight which samples suffered on treatment. He also looked at the response of a range of 16 freshly quarried Jurassic limestones from the Cotswold Hills near Cheltenham, UK, and found that there was another significant correlation of $S = 0.72$ between the water absorption capacity and the loss of weight which samples suffered on treatment. A higher correlation (0.79) was found between the bulk specific gravity (measured by the paraffin wax method) and the loss of weight. The content of non-calcareous material (the insoluble residue) in the limestones, which might at first sight have been expected to affect the resistance (see, for example, Sparks, 1949 on the English Chalk) showed no correlation: $S = -0.11$ (Table 4.12).

Some further experimental information on the response of different rock types is provided by Goudie and Viles (1995). They subjected six sedimentary rocks to a Negev cycle using saturated solutions of sodium carbonate and sodium sulphate. There was a clear relationship between the water absorption capacity, the amount of salt taken up in the rock pores and the amount of breakdown that occurred after 100 cycles (Table 4.13).

What is evident from all these studies is that resistance to salt attack can in a general way be related to the amount of moisture and salt that a rock can

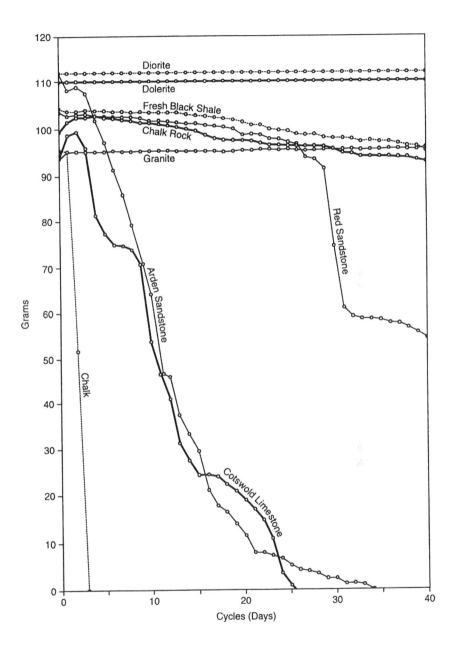

Figure 4.7. Changes in the weight of selected rock types on treatment with sodium sulphate. From Goudie (1974: figure 3). Reproduced with permission

Table 4.11. Rock type, water absorption capacities, and resistance to Na_2SO_4 crystallisation

Rock type capacity	Water absorption capacity (%)	Remaining after 30 cycles of salt crystallisation with Na_2SO_4 (%)	No. of blocks
Diorite	0.33	99.80	2
Dolerite	0.21	99.72	2
Granite	0.14	99.68	2
Chalk rock	4.99	95.89	2
Calcrete	2.62	94.60	1
Shale	0.65	93.67	20
Gneiss	0.22	93.19	2
Calcrete	1.67	92.80	1
Schist	0.57	90.70	2
Old Red Sandstone	3.38	88.02	2
Arden Grit	4.60	61.72	2
Cotswold Limestone	6.20	50.35	23
Silica-cemented sandstone	3.30	11.64	3
Arden Sandstone	8.97	2.44	3
Arden Shale	7.10	2.28	2
Calcrete	13.08	0.00	1
Chalk	19.89	0.00	5

Source: Goudie (1974: table 4). Reproduced with permission.

absorb. It is also evident that sedimentary rocks with high water absorption capacities show rapid rates of breakdown in the face of a range of salt weathering simulations, whereas igneous and metamorphic rocks are on the whole much more resistant. That is not to say, however, that salt weathering experiments have been totally ineffectual in causing the breakdown of rocks such as granites, for there are experiments that prove otherwise (e.g. Pedro, 1957a; Kwaad, 1970). There is also abundant field evidence that many types of rock are susceptible to salt attack and Table 4.14 lists some examples of studies that have been undertaken on non-sedimentary rock types. There is considerable scope for a more detailed consideration of the effects of rock properties on their susceptibility to salt attack, and a need to consider such parameters as their pore size distribution (Knöfel et al, 1987).

EXPERIMENTAL SIMULATIONS OF CO-ASSOCIATIONS OF SALT WEATHERING AND OTHER WEATHERING PROCESSES

Several studies have been carried out to investigate the linked action of salt crystallisation and hydration and other processes of materials breakdown, most notably frost action. In real situations salt weathering may act before, after, or at the same time as a range of other processes, and it is of more than

Table 4.12. Resistance of Cotswold limestone samples in relation to selected properties

Sample No.	Remaining after 30 cycles of Na_2SO_4 salt crystallisation (%)	Water absorption capacity (%)	Bulk specific gravity	Insoluble residue (%)
CL27	73.64	3.21	2.35	5.49
CL28	81.24	3.10	2.44	2.20
CL29	56.85	5.04	2.23	6.89
CL30	81.54	2.21	2.46	6.04
CL31	90.60	4.08	2.37	3.30
CL32	46.21	9.68	1.95	12.49
CL33	31.85	8.49	2.15	2.60
CL34	40.84	6.02	2.24	1.48
CL35	38.85	3.90	2.32	2.40
CL36	81.53	3.63	2.37	1.62
CL37	25.25	11.87	2.09	2.08
CL38	43.44	4.88	2.25	21.56
CL39	6.62	11.52	1.86	40.56
CL40	43.12	6.11	2.26	3.14
CL41	75.64	3.33	2.46	4.45
CL42	83.43	5.39	2.39	1.11

Source: Goudie (1974: table 8).

just academic interest to look at how such co-associations work. As with experiments on salt weathering alone there has been little standardisation of experimental design in such studies. Early laboratory simulations by Goudie (1974) and Williams and Robinson (1981) indicated that some rocks disintegrate more rapidly when they are frozen after soaking in a salt solution rather than pure water, whereas the experiment of McGreevy (1982) indicated that salts may reduce or even prevent frost weathering.

A series of experiments on salt and frost weathering has been carried out by Jerwood and co-workers (discussed in Chapter 5) using chalk, following observations of the intense breakdown of chalk shore platforms in southern England after harsh winter conditions.

More recently, Wessman (1996) has carried out tests on frost and salt weathering (using sodium chloride and $Na_2SO_4 \cdot 10H_2O$ at a range of different concentrations) on Swedish sandstones, limestone and granite. Two types of tests were used: the first measured weight loss after a number of freeze–thaw cycles with immersion throughout; and the second measured dilatation after one freeze–thaw cycle. Stone samples of $3 \times 3 \times 12$ cm dimensions were used. Those of limestone and sandstone were subjected, in the scaling test, to 20 freeze–thaw cycles (except in the case of limestone in sodium chloride where only four cycles were used because of the high rate of breakdown), whereas

Table 4.13. Rock properties and breakdown

Rock type	Water absorption capacity (%)	Increase in weight from salt uptake (%)		Degree of breakdown after 100 cycles expressed as debris liberation as % of original weight	
		Na_2CO_3	Na_2SO_4	Na_2CO_3	Na_2SO_4
Chalk[1]	15.05	13.89	13.26	64.00	72.87
Ketton Stone[2]	7.30	5.85	5.99	4.23	3.43
Bath Stone[3]	10.30	10.98	10.17	7.43	16.32
York Stone[4]	2.19	1.99	1.60	0.098	0.25
Portland Stone[5]	5.44	5.43	5.91	0.61	2.49
Portuguese Stone[6]	2.73	2.48	2.73	0.34	7.22

[1] Chalk. A soft white Cretaceous limestone from the Upper Chalk of southern England, with a water absorption capacity (WAC) of 15.05% and a fine-grained texture.
[2] Ketton Stone. A soft Jurassic oolite limestone from the Inferior Oolite (Lincolnshire Limestone) of the East Midlands of England, with a WAC of 7.30% and large ooids of more than 500 μm in diameter, held together by meniscal micritic cement.
[3] Bath Stone. A soft Jurassic oolite limestone from the Great Oolite Series of south-west England, with a WAC of 10.30%. The ooids are generally >500 μm in diameter, and shell fragments are also common.
[4] York Stone. A buff, fine-grained, Carboniferous sandstone from the Coal Measures of West Yorkshire, northern England, with a WAC of 2.19%.
[5] Portland Stone. A creamy white, close-grained, shelly, fine-grained oolite Jurassic limestone from the Isle of Portland, Dorset, southern England. The ooids are generally about 200 μm in diameter and the WAC is 5.44%.
[6] Portuguese Stone. A creamy brown, coarse-grained upper Jurassic limestone from near Fatima in Central Portugal, containing fossil fragments, and with a WAC of 2.73%.
Source: from data in Goudie and Viles (1995).

those of granite were subjected to 35 freeze–thaw cycles. The freeze–thaw regime used in the scaling tests cycled from +20 to −15°C and back over a period of 24 hours.

Within the scaling tests, Wessman (1996) found the following results. Both limestone and sandstone showed a higher degree of weight loss in the sodium chloride tests than with sodium sulphate, with particularly large differences between the two limestone tests where breakdown differed by a factor of 100. Limestone appeared to be sensitive to breakdown in pure water and 4% sodium chloride solution, and much less sensitive to breakdown in solutions of intermediate strength. Two types of sandstone were used in the tests, and the more porous of the two proved less prone to breakdown in the sodium chloride tests than the denser sandstone. The relatively non-porous granite showed little breakdown, even after 35 cycles, and the weight loss was greatest in specimens treated with sodium sulphate, then pure water, and lastly sodium chloride. In the dilatation tests, very different results were found. For the

Table 4.14. Non-sedimentary rock types for which there are field observations of salt attack

Location	Rock type	Source
Devon, UK	Greenschist	Mottershead and Pye (1994)
Namibia	Gneisses, granites	Lageat (1994)
Iran	Green tuff and andesite	Beaumont (1968)
Dahomey	Granite	Tricart (1960)
Death Valley, USA	Rhyolite, quartz	Goudie and Day (1980)
Tenerife	Basalt	Höllerman (1975)
Antarctica	Dolerite	Miotke and Hodenberg (1980)
Karakorams, Pakistan	Granite, schist, slates	Goudie (1984)
Kenya	Basalt	Smith and McAlister (1986)
South Australia	Granite	Bradley et al (1978)
Queensland	Granite, quartz	Coleman et al (1966)
Northern Territory (Australia)	Quartize	Dunn (1915)
Liberia	Dolerite	Tricart (1962)
Basin and Range Province, USA	Granodiorite, rhyolite, basalt, monzonite, gabbro, tuff and syenite	Kirchner (1996)

sandstones, which were the only samples to show any appreciable dilatation response, dilatation was minimal at saturations below 0.9‰, but increased dramatically at 1‰. The more porous sandstone showed higher dilatations than the denser sandstone (which is the opposite of the scaling tests), except in pure water. Sodium sulphate caused slightly greater dilatation than sodium chloride. Both limestone and granite showed little change because of their low porosities. As Wessman acknowledges, the two tests are very different and represent different destruction mechanisms (Wessman, 1996: 571).

EXPERIMENTAL SIMULATIONS OF SULPHATION

A different approach has been taken in studies of sulphation, where gaseous pollutants are introduced into a specially designed flow chamber to react with specimens. As with the experiments discussed in the previous sections, there has been little standardisation of experimental design. Table 4.15 lists some of the many experiments carried out by a range of laboratories. In most cases a range of gases is introduced into the chamber (often sulphur dioxide, nitrogen oxide and nitrogen dioxide) at various concentrations, at a pre-set relative humidity (usually greater than 80%), sometimes in the presence of a range of catalysts and oxidants. In Britain, many experiments of this type utilise two commonly used British building limestones, Portland stone and Monk's Park stone; in other countries important local stones may be used, such as studies on travertines in Turkey (Götürk et al, 1993).

The experiment of Haneef et al (1992) indicates the type of results that can be obtained from such laboratory simulations. As shown in Table 4.15, four stone types were used and exposed to realistic concentrations (10 ppmv) of sulphur dioxide, nitrogen oxide or nitrogen dioxide. After 30 days only small weight changes were observed, except in the additional presence of ozone, when both Portland limestone and Mansfield sandstone showed a greater than 10 mg weight gain in the nitrogen oxide and nitrogen dioxide runs. Weight gains were shown by all wetted samples under the sulphur dioxide regime (with Portland limestone gaining 60 mg), but weight losses under the nitrogen oxide and nitrogen dioxide regimes. Retained ion concentrations are shown in Table 4.16, which indicates the general retention of sulphates and nitrates under most experimental conditions. Calculations of calcium yields in the experiments show that the extent of chemical reaction of the wetted stones with sulphur dioxide and nitrogen dioxide is up to four times that of dry stones.

Experimental simulations of the sulphation process and other linked processes have been extremely important in improving our understanding of the reaction between stones rich in calcium carbonate and polluted atmospheres.

EXPERIMENTAL SIMULATIONS OF THE CHEMICAL ACTION OF SALTS

As well as the largely physical processes of decay involved in salt crystallisation and hydration, salt solutions may have a series of chemical effects on materials. Salt solutions possess a wide range of pH values and may produce etching and the transformation of a range of minerals. Gillott (1979) carried out pioneering experiments on the effects of de-icing agents and sulphate solutions on concrete aggregate, illustrating the chemical as well as physical nature of the decay. Small rock cylinders of five different limestones were immersed in pure water, as well as 5% and saturated solutions of sodium chloride and calcium chloride and 4% and saturated solutions of magnesium sulphate for more than two years at 20°C, with regular observations taken. Small samples of the limestones were removed after a few days and examined under a scanning electron microcope, revealing common solution effects on calcite crystals, coupled with the rarer precipitation of new minerals. Cleavages, grain boundaries and weaknesses of all kinds were preferentially attacked. All salts seemed to produce etching, although calcium chloride was the only salt to produce notable etch pitting.

As scanning electron microscope observations are increasingly made of samples before, during and after a range of experimental weathering simulations it should be possible to expand this type of work to investigate the initial stages of weathering at the mineral grain and crystal scale.

Table 4.15. Some selected experimental simulations of sulphation and similar processes

Reference	Rock types used	Measurements taken	Experimental design
Götürk et al (1993)	Travertine (in powdered form and 30 × 40 × 2 mm slabs)	% weight gain; XRD of reaction products; SEM; determination of sulphate contents	To investigate: (1) effect of SO_2 concentrations of 750–4000 ppm; (2) effect of RH from 0 to 100%; (3) effect of temperatures from 15 to 70°C. Tests run for 80–100 h
Haneef et al (1992)	Portland and Massangis limestones, White Mansfield dolomitic sandstone, Pentelic marble 50 × 30 × 5 mm slabs	Weight change; sulphate and nitrate contents by high performance liquid chromatography	To investigate: impacts of sulphur dioxide, nitrogen oxide and nitrogen dioxide singly and with ozone on dry and wetted stones 84% RH, 294 K temperature, 10 ppmv of all gases, realistic wetting regime, 30 days each run
Kirkitsos and Sikiotis (1995)	Pentelic marble, Portland limestone and Baumberger sandstone crushed to grains of 0.2–1 mm size	Ion chromatography of nitrates	To investigate: influence of nitric acid on various stones under different grain sizes, flow-rates and concentrations RH 0–100%; HNO_3 concentration 800–3200 μg m^{-3}
Lal Gauri et al (1989)	Georgia marble slabs (dimensions not given)	Light and SEM microscopy; porosity by water absorption	To calculate reaction rate constant for marble sulphation 20°C, 100% RH; 10 ppm sulphur dioxide
Hutchinson et al (1992)	Portland and Monks Park limestones (sample nature and dimensions not given)	SEM/energy dispersive analysis of X-rays/electron probe micro-analysis and optical microscopy	To compare the action of HCl and sulphur dioxide on limestone 95% RH, 25 ppm HCl and sulphur dioxide; 25 days for HCl run; 50 for sulphur dioxide

Table 4.16. Retained ion contents (mg) on dry and wet stones exposed to the pollutant gases in the presence and absence of ozone for 30 days, from the experiment of Haneef et al (1992)

Stone	Exposure regime					
	NO (nitrate)	NO_2 (nitrate)	SO_2 (sulphate)	$NO + O_3$ (nitrate)	$NO_2 + O_3$ (nitrate)	$SO_2 + O_3$ (sulphate)
Pentelic marble (dry)	5.3	6.3	4.9	24.7	50.1	49
Mansfield sandstone (dry)	13.9	12.0	13.7	16.2	85.6	75.3
Massangis limestone (dry)	5.2	17.8	9.8	100.8	46.4	63.9
Portland limestone (dry)	14.7	6.8	17.5	172.8	88.8	67.9
Pentelic marble (wet)	none	13.0	17.8	15.1	17.0	52.8
Mansfield sandstone (wet)	1.8	24.1	36.2	52.6	23.9	63.1
Massangis limestone (wet)	0.4	17.4	28.7	26.8	16.7	87.5
Portland limestone (wet)	3.8	19.5	51.2	45.6	28.1	129.5

FIELD EXPERIMENTS ON SALT WEATHERING IN CO-ASSOCIATION WITH OTHER PROCESSES

Both building materials scientists and geomorphologists have commonly used field experimentation, or exposure trials, as an adjunct method to discover more about the rate and nature of a variety of weathering processes. Such tests have the advantage of using realistic environmental conditions and any dangers of acceleration of weathering cycles are also avoided. However, there is a much lower degree of control over the progress of weathering in such tests, and there may be a bewildering array of variables and processes to consider.

The basic protocol of such tests is to leave samples of materials in the environment for a suitable length of time, monitoring the environmental conditions during exposure and measuring the change in material properties at the end of the test. As with laboratory simulations, however, there is wide variety in the detailed experimental design used by such tests which limit their comparability. It is also often hard to compare the results from such tests with the results from controlled laboratory simulations. Goudie et al (in press), for example, left small blocks of Bath stone out on desert surfaces within the Namib desert for two years (Figure 4.8) and used scanning electron microscope and chemical analyses to investigate the degree and causes of

Figure 4.8. (a) Bath Stone slab after two years of exposure in the Namib desert showing little visible deterioration (site no. 7). (b) Bath Stone slab after two years of exposure in the Namib desert showing complete disintegration (site no. 8) (photographs by H. A. Viles)

decay. A wide range of national and international monitoring networks using such natural stone sensors have also been set up to look at the influence of polluted atmospheres and natural processes on calcareous stone decay, such as the National Materials Exposure Programme in Britain (from 1987 to 1995; Butlin et al, 1992), the International Materials Exposure Programme co-ordinated by UNECE, and the EUREKA co-ordinated EUROMARBLE exposure programme which began in 1992 (Simon and Snethlage, 1996). Many of these exposure programmes use small blocks of stone mounted on freely rotating carousels at sites exposed to rain, and also at sites sheltered from the rain. A range of pre- and post-exposure measurements are taken, ranging from observations of surface roughness, the content of soluble salts, weight and colour change. Comparisons can be made between the weathering behaviour of similar materials in a wide range of environments under natural conditions using such monitoring networks.

CONCLUSIONS

This chapter has shown how important the physical and chemical characteristics of a material are to its sensitivity to salt weathering. We have also shown the value and scope of experimental weathering simulations of various kinds in furthering our knowledge of the rate and nature of the salt weathering processes. However, it is also apparent that the unco-ordinated nature of much experimentation makes inter-comparisons difficult and seriously hinders scientific progress in this field. It would be exciting to see increased co-ordination between different types of experiments so that we may learn more about the chemical and physical effects of salts in co-association with a range of other weathering processes under differing environmental conditions.

5 Mechanisms of Salt Attack

INTRODUCTION

Salts attack rocks and building materials in a range of ways which can, conventionally, be divided into two main groups: those involving a predominantly physical change in the state of the material and those involving some change in the chemical state of the material (Table 5.1). Despite such a classification into physical and chemical changes, the chemical behaviour of soluble salts is at the heart of salt weathering and must be understood before a real appreciation of how salts cause damage to materials can be gained. Two aspects need to be appreciated: firstly, the thermodynamics of the reactions involved (i.e. how likely, on energy grounds, a reaction is to occur); and, secondly, the kinetics of reactions (i.e. given that a reaction is thermodynamically possible, how fast it will occur). As we are dealing with complex ionic compounds, often in mixed solutions, such an understanding is by no means straightforward. Although laboratory experiments have proved highly useful in understanding the thermodynamics and kinetics of many reactions, geochemical modelling is also increasingly needed to aid prediction and understanding.

CRYSTALLISATION OF SALTS

The most cited cause of salt weathering is generally the process of salt crystal growth from solutions in rock pores and cracks (Figure 5.1) and the theory of

Table 5.1. Mechanisms of salt attack

Physical changes
  Crystallisation
  Hydration
  Thermal expansion
  Electrical slaking double layer effects associated with hygroscopicity

Chemical changes
  Silica mobilisation under alkaline conditions
  Etching of calcite under acid conditions
  Changes to concrete mineralogy
  Corrosion of incorporated iron and steel
  Moisture-related chemical weathering associated with hygroscopicity
  Gypsum/silicate replacement

Figure 5.1.   A close up illustration of the weathering of boulders on fans entering the salty zone of the Death Valley Playa near Badwater (photograph by A. S. Goudie)

this process has been reviewed by Evans (1970). Data on crystallisation pressures are given in Winkler and Singer (1972).

Various mechanisms can cause crystal growth to occur. For example, some salts rapidly decrease in solubility as temperatures fall. This is particularly true of sodium sulphate, sodium carbonate, magnesium sulphate and sodium nitrate, though not to the same degree of calcium sulphate and sodium chloride (Table 5.2). Thus nocturnal cooling could cause salt crystallisation to occur. Such a crystallisation of a salt solution on a temperature fall affects a much larger volume of salt per unit time than crystallisation induced by evaporation, which is a more gradual process (Kwaad, 1970: 79).

Nevertheless, evaporation does help to create saturated solutions from which crystallisation can occur, and when this happens highly soluble salts will produce large volumes of crystals (Figure 5.2). In this context it is important to note that of the common salts, gypsum is much less soluble than many of the others, and that less crystalline material will be available in a given volume of solution to cause rock disruption.

The crystal habit of the various salts may also affect their power to cause rock breakdown. For instance, the needle-shaped habit of sodium sulphate crystals (mirabilite) might tend to increase their disruptive capability.

In some bedrock masses, and in some sediments, such may be the accumulation of crystalline material that ground heaving occurs, producing mounds and pseudo-anticlines (e.g. Watts, 1977; Watson, 1982).

Figure 5.2. Volume expansion and fragmentation in a limestone in Ras Al Khaimah, caused by the build up of salts in weaknesses in the rock. The thickness of the block is approximately 12 cm (photograph by A. S. Goudie)

Air humidity is a highly important control of the effectiveness of salt crystallisation, for a salt can crystallise only when the ambient relative humidity is lower than the equilibrium relative humidity of the saturated salt solution. If that is the case on a rock surface then the salt will crystallise and cause decay. The equilibrium relative humidities of different salts vary considerably, and those with low values will be prone to dissolution in humid air. The equilibrium relative humidities of hydrated sodium carbonate and sodium sulphate are high, whereas those of sodium chloride, sodium nitrate and calcium chloride are relatively lower (Arnold, 1981; Table 5.2).

How precisely does salt crystallisation operate? This is an important issue for, as Lewin (1990: 59) has pointed out, research has often been characterised by weak suppositions rather than by strong theory.

A substantial literature exists describing the manifestations of salt decay in a variety of individual monuments, buildings and other structures.

Many of these reports are qualitative (i.e., anecdotal) in nature; a relative few are quantitative. The only concept that has been advanced in "explanation" of those observations is the suggestion that there exists a phenomenon termed "crystallization pressure". According to this interpretation, a growing crystal that has become confined within a pore in stone (due to its growth) continues to grow even after there is no longer any void space, and in so doing exerts a pressure against the confining walls which can ultimately lead to the disruption of the host by the guest.

Table 5.2. Solubility of some common salts in water

Salt	Solubility (as % at 0°C)	Solubility (as % at 35°C)	Solubility at 0°C (as a % of that at 35°C)
Na_2CO_3	6.54	32.90	19.88
$NaNO_3$	42.40	49.60	85.08
Na_2SO_4	4.76	33.40	14.25
NaCl	26.28	26.57	98.91
$MgSO_4$	18.00	29.30	61.43
$CaSO_4$	0.19	0.21	90.48

However, to say that an object has been damaged as a consequence of a mysterious internal crystallization pressure is not any more of an explanation than to say that it has been damaged by a tendency to decay. Indeed, the term is rather anthropomorphic, suggesting that the crystal feels a "need" to grow. If the concept of crystallization pressure is to contribute to our understanding of stone decay, the origin, nature and parameters of that putative pressure must be expressible in terms of the laws of physics and chemistry.

Moreover, as Lewin (1981: 121) has pointed out elsewhere

However, it is not immediately evident why the deposition of a solute from a solution into a pore or crack at the surface of a solid should damage the latter. Consider, for example, the evaporation of a sodium chloride solution at a stone surface. When, as a consequence of the escape of water vapour, the solution reaches saturation, it contains at ordinary temperatures about 26 percent solid matter by weight. Hence, a pore filled with such a solution can have only about one-quarter of its volume taken up by the residue left when evaporation is complete. Each repeated imbibition of salt solution can be expected to reduce the remaining free volume of the pore by one quarter of that value, until the pore is filled with deposited solute. But there is no analogy in this process to the expansive force that develops when water filling a pore transforms into ice or when certain types of solid phases filling a pore recrystallize into higher hydrates (as, for example, when sodium sulphate (Na_2SO_4) transforms at high humidity into $Na_2SO_4 \cdot 10H_2O$).

Thus, as Lewin points out, if pore solutions evaporate and deposit salt crystals, pore spaces will in due course become filled. To go beyond that stage crystals would have to exert themselves to make space against the opposing force of the pore walls and the crystal mass that is already filling the space between those walls. Lewin (1990: 63) believes that it is the release of free energy during the spontaneous deposition of crystals from a sufficiently super-saturated solution that supplies the energy to compress both themselves and the stone matrix to a magnitude sufficient to cause mechanical failure.

Following Correns (1949), Winkler and Singer (1972) have calculated crystallisation pressures using the following formula:

$$P_c = \left(\frac{RT}{V_c}\right)\ln C/C_s$$

Where P_c = pressure exerted by crystal growth; R = gas constant of the ideal gas law, $8.3145\,\mathrm{J\,mol^{-1}\,K^{-1}}$. T = temperature (K); V_c = molecular volume of the solid salt; C = actual concentration of the solute during crystallisation; and C_s = concentration of the solute at saturation.

Some of the data calculated on this basis have been determined for particular salts, as listed in Table 5.3.

As Cooke and Gibbs (no date: 62) have pointed out, P_c in this formula is not necessarily the "pressure exerted by crystal growth." They explain it thus

If supersaturation develops in an unconstrained situation, crystallization occurs but there is no pressure exerted. If a crystal starts to grow in a confined space in a porous solid, a stress will build up as the volume change starts to deform the solid elastically. This stress will increase with increasing deformation until it is relieved either by plastic deformation (of a ductile solid) or cracking (of a brittle solid such as stone), or it rises to the value P_c. The point about Correns' equation is that, for supersaturations that can readily develop, P_c is very large and is probably not attained before the critical stress for internal cracking of porous stones is reached. The internal stresses actually developed at the onset of cracking will then be directly proportional to the amount of crystallization that has taken place, the crystal molar volume V_c and the elastic moduli of the stone.

It follows that the theoretical crystallization pressures P_c for various salts do not necessarily rank them in order of propensity for causing damage. So long as they all exceed the stresses required to cause internal cracking, the crystal characteristic that should be an indicator of damage propensity is the crystallization volume, V_c. This accords with empirical observation.

There are certain question marks over the uncritical acceptance of the Correns' equation. As C. Price (pers. comm. 1997) remarked, "How realistic is it to think that one can get significant degrees of supersaturation in a dirty piece of stone, with any number of possible nucleation sites? For example, won't the receding front of solution leave behind a supply of unstressed crystals that can act as nuclei?"

Fitzner and Snethlage (1982) developed an equation which considers crystallisation pressures more explicitly in relation to porous materials. Fitzner and Snethlage (1982), following the work of Everett (1961), approach the theoretical problem of crystallisation pressures in porous media in terms of an analogy with the thermodynamics of freezing of solutions in porous materials. In their work

$$P = 2\sigma\left(\frac{1}{rR}\right)$$

where P = crystallisation pressure; σ = ionic interfacial tension of the salt solution; r = radii of small pores; and R = radii of large pores.

Table 5.3. Crystallisation pressures for some salts (after Winkler and Singer, 1972)

Salt	Chemical formula	Density (g/cm³)	Molecular weight (g/mol)	Molar volume (cm³/mol)	Crystallisation pressure (atm)					
					$C/C_s = 2$ 0°C	$C/C_s = 50$ 0°C	$C/C_s = 2$ 50°C	$C/C_s = 50$ 50°C	$C/C_s = 10$ 0°C	$C/C_s = 10$ 50°C
Anhydrite	$CaSO_4$	2.96	136.14	46.00	335	1900	398	2262	1120	1325
Bischofite	$MgCl_2 \cdot 6H_2O$	1.57	203.31	129.50	119	675	142	803	397	470
Dodekhaydrate	$MgSO_4 \cdot 12H_2O$	1.45	336.00	231.90	67	378	80	450	222	264
Epsomite	$MgSO_4 \cdot 7H_2O$	1.68	246.48	147.00	105	595	125	708	350	415
Gypsum	$CaSO_4 \cdot 2H_2O$	2.32	127.00	54.80	282	1595	334	1900	938	1110
Halite	$NaCl$	2.17	58.54	27.85	554	3135	654	3737	1845	2190
Heptahydrite	$Na_2CO_3 \cdot 7H_2O$	1.51	232.10	153.80	100	568	119	677	334	365

Mineral	Formula							
Hexahydrite	$MgSO_4 \cdot 6H_2O$	1.75	228.00	130.10	118	141	395	469
Kieserite	$MgSO_4 \cdot H_2O$	2.45	138.39	56.55	671	300	910	1079
Mirabilite	$Na_2SO_4 \cdot 10H_2O$	1.46	322.19	220.00	272 / 1543	324 / 1840	234	277
Natron	$Na_2CO_3 \cdot 10H_2O$	1.44	286.14	198.70	72 / 397	83 / 473	259	308
Tachyhydrite	$2MgCl_2 \cdot CaCl_2 \cdot 12H_2O$	1.66	514.40	309.50	78 / 440	92 / 524	166	198
Thenardite	Na_2SO_4	2.68	142.04	53.0	50 / 282	59 / 336	970	1150
Thermonatrite	$Na_2CO_3 \cdot H_2O$	2.25	124.00	55.0	292 / 1650 / 280 / 1590	345 / 1965 / 333 / 1891	935	1109

1 atm = 0.1013 MPa.

Table 5.4.   Material strengths and salt pressures

(A) Tensile strengths of materials in MPa	
Extremely high strength rocks[1]	>10
Very high strength rocks[1]	3–10
High strength rocks[1]	1–3
Medium strength rocks[1]	0.3–1
Low strength rock[1]	0.1–0.3
Very low strength rocks[1]	0.03–0.1
Extra low strength rocks[1]	<0.03
Concretes (typical values)[2]	2–4
(B) Pressures produced by salt processes in MPa	
Expansion of steel reinforcements on rusting[2]	Up to 30
Crystallisation pressures of gypsum[3]	28.2–19.0
Crystallisation pressures of halite[3]	5.54–373.7
Crystallisation pressures of thenardite[3]	29.2–196.5
Hydration pressures of gypsum[4]	Up to 254
Hydration pressures of $MgSO_4$[4]	Up to 42
Hydration pressures of Na_2SO_4[4]	Up to 48

Sources: [1]Bell (1992); [2]Murdock et al (1991); [3]Winkler and Singer (1972); and [4]Kirchner (1996).

Crystallisation, according to this theoretical approach, occurs first in large pores and only moves into small pores once the larger ones are full. Until that point the small pores act as supply reservoirs for salt solutions. Stones with a high volume of large pores and a higher volume of smaller pores are seen to be particularly sensitive to salt weathering and frost weathering.

The pressures produced both by salt crystallisation and by salt hydration (Knacke and Erdberg, 1975) appear to exceed comfortably the tensile strength of most rocks and concrete as the data in Table 5.4 show. The materials typically have strengths of the order of 0.1–10 MPa, whereas the crystallisation and hydration pressures shown in Tables 5.3 and 5.6 and summarised in Table 5.4B range widely, but are frequently of the order of some tens or hundreds of MPa.

The presence of crystallising soluble salts does not invariably cause decay because salt crystallisation can be either disruptive or cause cementation depending on the pore structure of the material and the crystallisation pressures that develop with particular pore size distributions. However, cementation may lead indirectly to decay by fundamentally altering the surface characteristics of a material compared with those at depth. Rossi-Manaresi and Tucci (1991) observe that high crystallisation pressures (greater than $350\,N\,cm^{-2}$) are associated with stones where the stone has a structure that has a substantial percentage of small ($<0.5\,\mu m$) pores as well as pores of larger size ($>5\,\mu m$). When the percentage of small pores is much lower, the crystallisation pressures can only build up to lower values, which are not sufficient to lead to mechanical failure. In such cases salt crystallisation may cause surface hardening rather than disintegration.

Situations can become very complex when mixtures of salts are present in solutions, and several empirical and theoretical studies have addressed this issue. La Iglesia et al (1994) investigated the interaction of solutions on buildings stones in the Cathedral of Toledo, Spain and concluded that runoff waters were sulphate solutions, and that the following salts would precipitate out in this order: gypsum, bloedite, epsomite, kieserite, mirabilite and thenardite. Using the theoretical approach of Fitzner and Snethlage (1982) they predicted that gypsum would produce the greatest crystallisation pressure, followed by anhydrite, mirabilite, epsomite, bloedite, thenardite and kieserite. In a different setting, Zehnder (1996) considers the order in which salts will precipitate out of a mixed solution in a zone of rising damp. As a dilute solution rises up a wall due to capillary action it becomes subject to evaporation and concentration and salts precipitate in an order relating to their solubilities. Thus in a solution containing sodium, potassium, magnesium, calcium, chloride, nitrate and sulphate, weakly to moderately soluble salts will precipitate out first (e.g. gypsum, epsomite and soda nitre). As these have low hygroscopicities they can easily crystallise even in a humid climate, producing much deterioration. Higher up the wall more soluble salts precipitate out (e.g. nitratine, halite, nitromagnesite and nitrocalcite). These cause less damage because they are more hygroscopic and thus crystallise less, except under dry conditions such as in an arid environment or heated building. Zehnder (1996) has also shown that gypsum accumulates at the top of zones of rising damp. He explains this using thermodynamic calculations on salt mixtures which show that the solubility of gypsum increases in the presence of many other salts and thus its deliquescence humidity is lowered considerably.

HYDRATION

Certain common salts hydrate and dehydrate relatively easily in response to changes in temperature and humidity. As a change of phase takes place to the hydrated form, water is absorbed. This increases the volumes of the salt and thus develops pressure against pore walls. The volume increases for some common salts are given in Table 5.5, where it can be seen that the change may be appreciable, with sodium carbonate and sodium sulphate both undergoing a volume change in excess of 300% as they hydrate. As with salt crystallisation there is still some mystery attached to this mechanism. In many cases the anhydrous salt will have formed in the first place from a salt solution (or from the hydrated salt) within the pores, and so it ought therefore to have room for it to expand back. This needs further investigation.

For some salts a change of phase may occur at the sorts of temperatures encountered widely in nature; sodium sulphate's transition temperature is 32.4°C for a pure solution, and falls to 17.9°C in a sodium chloride saturated

Table 5.5. Hydration volume increase for selected common salts

Salt	Molecular weight	Hydrate	Formula weight of hydrate	Density	Density of hydrate	Volume change (%)
Na_2CO_3	106.00	$Na_2CO_3 \cdot 10H_2O$	286.16	2.53	1.44	374.7
Na_2SO_4	142.00	$Na_2SO_4 \cdot 10H_2O$	322.20	2.68	1.46	315.0
$CaCl_2$	110.99	$CaCl_2 \cdot 2H_2O$	147.03	2.15	0.84	241.1
$MgSO_4$	120.37	$MgSO_4 \cdot 7H_2O$	246.48	2.66	1.68	223.2
$CaSO_4$	136.14	$CaSO_4 \cdot 2H_2O$	172.17	2.61	2.32	42.3

Source: modified after Goudie (1977: table VII).

environment. Moreover, for some salts the transition may be rapid. At 39°C the transition from thenardite (Na_2SO_4) to mirabilite ($Na_2SO_4 \cdot 10H_2O$) may take no longer than 20 minutes (Mortensen, 1933).

An early experimental verification of the power of hydration in porous building materials was provided by Bonnell and Nottage (1939: 20). They developed a simple, robust mould under compression, altered its temperature and measured volume changes for different mixes of salts (sodium sulphate and magnesium sulphate) and sand. From these experiments they concluded

> The anhydrous salt (or the lower hydrate) may be hydrated (or further hydrated) even against moderately high stresses. Since these stresses are well above the tensile strength of normal porous building materials, it is clear that, if hydration is allowed to take place within the pores of such a building material, the salt may possibly, during such hydration, exert a sufficient force to bring about disintegration of the material.

Winkler and Wilhelm (1970) have calculated the hydration pressures of some important common salts (Table 5.6) at different temperatures and relative humidities, and find that the greatest hydration pressures (maximum value 2190 atm at 0°C and 100% relative humidity) occur when anhydrite changes into gypsum. This is in excess of the crystallisation pressure of ice at $-22°C$, and is in excess of the pressure required to exceed the tensile strength of rocks.

To calculate the hydration pressures listed in Table 5.6, we can follow Mortensen (1933). He proposed the following equation:

$$P = \frac{RT}{V} \ln \left(\frac{P_1}{P_2} \right)$$

where P = hydration pressure ($N\,mm^{-2}$); T = absolute temperature (K); V = molar volume of water of crystallisation ($cm^3\,mol^{-1}$); R = gas constant; and p_1 = vapour pressure of water at the temperature T (mmHg) and p_2 = dissociation pressure of the hydrate at T (mmHg).

Winkler and Wilhelm (1970) used Mortensen's idea and extended it. For the calculation of hydration pressure, they developed the following equation:

Table 5.6. Hydration pressure (in atm) of some common salts versus temperature and relative humidity (after Winkler and Wilhelm, 1970)

Relative humidity	Temperature				Relative humidity	Temperature			
(%)	0°C	20°C	40°C	60°C		0°C	10°C	20°C	30°C
CaSO$_4$·½H$_2$O to CaSO$_4$·2H$_2$O Bassanite to gypsum					Na$_2$CO$_3$·H$_2$O to Na$_2$CO$_3$·7H$_2$O Thermonatrite to heptahydrite				
100	2190	1755	1350	926	100	938	770	611	430
90	2000	1571	1158	724	90	799	620	457	276
80	1820	1372	941	511	80	637	455	284	94
70	1600	1145	702	254	70	448	264	88	—
60	1375	884	422	0	60	243	46	—	—
50	1072	575	88	—					

	Temperature					Temperature			
MgSO$_4$·6H$_2$O to MgSO$_4$·7H$_2$O Hexahydrite to epsomite					Na$_2$SO$_4$ to Na$_2$SO$_4$·10H$_2$O Thenardite to mirabilite				
	10°C	20°C	30°C	40°C		20°C	25°C	31°C	32.4°C
	146	117	92	96	100	483	400	285	252
	312	103	77	69	90	331	246	135	93
	115	87	59	39	80	162	77	0	0
	97	68	40	5	75	66	0	—	—
	76	45	17	—					
	50	19	0	—					
	20	0	—	—					

	Temperature					Temperature			
Na$_2$CO$_3$·7H$_2$O N Na$_2$CO$_3$·10H$_2$O Heptahydrite to natron					MgSO$_4$·H$_2$O to MgSO$_4$·6H$_2$O Kieserite to hexahydrite				
	0°C	10°C	20°C	30°C		65.3°C			
	816	669	522	355	100	418			
	666	504	350	185	90	226			
	490	320	160	0	80	13			
	282	112	0	—	70	—			
	60	—	—	—	60	—			
					50	—			
					40	—			

$$P = \left(\frac{nRT}{V_h - V_a} \right) \times 2.3 \log \left(\frac{P_w}{P'_w} \right)$$

where P = hydration pressure (N mm^{-2}); n = number of moles of water gained during hydration to the next higher hydrate; R = gas content, (8.3145 J mol^{-1} K^{-1}); T = absolute temperature (K); V_h = volume of the hydrate, in cubic centimetres per gram-mole of hydrate salt; V_a=volume of

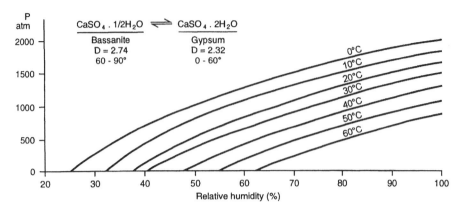

Figure 5.3. Plot of the hydration pressure from the hemihydrate form of calcium sulphate (bassanite) to the dihydrate form (gypsum) for various temperatures and relative humidities. Modified from Winkler and Wilhelm (1970) and reproduced with permission

original salt before hydration; P_w=vapour pressure of water in mm mercury at a given temperature and P'_w is the vapour pressure of the hydrated salt.

Winkler and Wilhelm (1970) also drew curves as three-dimensional contours within the fields of stability for each salt. On the x-axis is the relative humidity (%) of the atmosphere in contact with the salt, while on the y-axis is the pressure P (in atm). The contours are drawn at 10°C intervals between 0 and 60°C. They allow the exact theoretical determination of the maximum pressure exerted by a particular salt at a given humidity and temperature. It needs to be remembered that these curves are theoretical and present maximum values that might be attained under certain idealised conditions: closed pore systems, unlimited moisture access towards the hydrating crystals in the pores, etc. A plot of the hydration pressures created for the transition from bassanite ($CaSO_4 \cdot \frac{1}{2} H_2O$) to gypsum ($CaSO_4 \cdot 2H_2O$) is shown in Figure 5.3. The hydration pressures calculated for the bassanite–gypsum system have, however, been shown to be inaccurate by Kirchner (1995), who has also calculated the hydration pressures associated with the halite (NaCl) and hydrohalite system ($NaCl \cdot 2H_2O$). The Kirchner values are expressed as MPa rather than as atm (Table 5.7).

An important consideration in hydration is the speed at which the change of phase can take place. There is not a great deal of hard and sound information on this topic. Kwaad (1970) provided some data on this matter. To establish the hydration velocity of some hydrate-forming salts an hourly record was kept of their change in weight during 12 hours of exposure to 20°C and a relative humidity of 90% after salts had first been dried at 105°C. Particularly notable is the relatively rapid change in weight of magnesium sulphate as it transforms from $MgSO_4 \cdot H_2O$ to $MgSO_4 \cdot 6H_2O$. Sodium carbonate shows a

Table 5.7. Theoretical hydration pressures (in MPa) of common salts (recalculated after Winkler and Wilhelm, 1970: 568, 570 f). Numbers in parentheses behind pressures of 0 MPa indicate the relative humidity value below which no hydration takes place

$CaSO_4 \cdot \frac{1}{2}H_2O \rightarrow CaSO_4 \cdot 2H_2O$								$NaCl \rightarrow NaCl \cdot 2H_2O$				
Relative humidity (%)	Temperature °C							Temperature °C				
	0	10	20	30	40	50	60	−20	−15	−10	−5	0
100	254	231	207	190	159	136	112	66	60	54	48	42
95	245	221	196	179	148	124	100	59	53	47	41	35
90	235	211	186	168	136	112	88	51	46	39	33	27
85	224	200	174	156	124	100	75	44	38	32	25	19
80	213	188	162	144	111	86	61	36	30	23	17	10
75	201	176	149	130	97	72	46	27	21	14	7	1
70	188	162	135	116	83	57	31	18	12	5	0	0(74,7)
65	174	148	121	101	67	41	14					
60	159	133	105	84	50	23	0(61,2)					
55	143	166	87	66	31	4						
50	125	97	68	47	11	0(54,0)						

$Na_2CO_3 \cdot H_2O \rightarrow Na_2CO_3 \cdot 7H_2O$					$MgSO_4 \cdot H_2O \rightarrow MgSO_4 \cdot 6H_2O$	
100	94	77	61	43	Relative humidity (%)	Temperature (°C)
95	88	70	54	35		
90	80	62	46	28	100	42
85	72	53	37	20	95	32
80	64	46	28	9	90	23
75	55	36	19	0(74,5)	85	10
70	45	26	9		80	1
65	35	16	0(65,8)		79.2	0
60	24	5				
55	12	0(58,0)				
50	0(50,4)					

Continued

Table 5.7. *continued*

$Na_2CO_3 \cdot 7H_2O \rightarrow Na_2CO_3 \cdot 10H_2O$

100	82	67	52	36
95	76	59	44	27
90	67	50	35	19
85	58	41	26	9
80	49	32	16	0(80,1)
75	39	22	6	
70	28	11	0(72,3)	
65	17	0(65,1)		
60	6			
58	0(58,1)			

$MgSO_4 \cdot 6H_2O \rightarrow MgSO_4 \cdot 7H_2O$

Relative humidity (%)	Temperature (°C)			
	10	20	30	40
100	15	12	9	10
95	14	11	9	8
90	13	10	8	7
85	12	10	7	5
80	12	9	6	4
75	11	8	5	2
70	10	7	4	0.5
65	9	6	3	0(68.6)
60	8	5	2	
55	6	3	0.4	
50	5	2	0(53,4)	
45	4	0.4		
40	2	0(43,0)		
34.8	0			

$Na_2SO_4 \rightarrow Na_2SO_4 \cdot 10H_2O$

Relative humidity (%)	Temperature (°C)			
	20	25	31	32.4
100	48	40	29	25
95	41	32	23	18
90	33	25	14	9
85	25	16	4	1
80	16	8	0(83)	0(84,4)
75	7	0(76,2)		
71.5	0(71,3)			

$MgSO_4 \rightarrow MgSO_4 \cdot H_2O$

100	26	26
95	19	18
90	12	11
85	0(87)	3
83.2	0	

less marked and rapid change in weight, while sodium sulphate achieves no appreciable weight gain over the 12 hour period (Figure 5.4).

Figure 5.5 shows the result of a simple experiment undertaken by the present authors in which three salt types (sodium carbonate, sodium sulphate and magnesium sulphate) were dried for 24 hours at 105°C and were then placed in an environmental cabinet at 25°C and 90% relative humidity for 12 hours. Magnesium sulphate showed the greatest weight gain (16% in

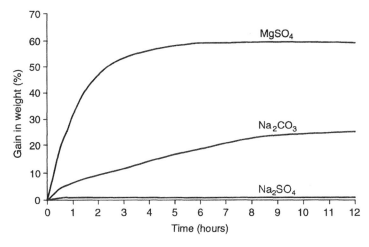

Figure 5.4. Hydration rate of Na_2SO_4, Na_2CO_3 and $MgSO_4$ at 20°C and at a relative humidity of 90%. Salts were dried to a constant weight at 105°C before the experiment. Modified from Kwaad (1970: figure 3)

12 hours) and was still absorbing more moisture when the cycle was terminated. Sodium carbonate absorbed about half as much moisture as magnesium sulphate and appeared to level off after about 10 hours. Sodium sulphate took up the least moisture (around 2%) and the rate of absorption levelled off after about six hours. These results broadly confirm the findings of Kwaad, though there are some small differences.

A comparable experiment was also undertaken to ascertain the speed with which weight gain took place in the context of blocks of York Stone which had been immersed in selected saturated solutions and then dried and subjected to a 12 hour cycle at 25°C and 90% relative humidity. All three treatments (Figure 5.6) showed a rapid weight gain in the first four hours, with the greatest increase in weight being achieved in this case by the sodium sulphate treated rock. In all three cases the weight gain can take place easily in the course of one daily cycle.

The number of occasions upon which rock surface temperatures cycle across the sorts of temperature thresholds associated with the change of phase from an hydrated to a dehydrated state is probably substantial in many desert areas. If we make the assumption that an air temperature of about 17°C translates into a rock surface temperature of about 32°C (the transition temperature for sodium carbonate and sodium sulphate), then that value is crossed daily on between five and nine months of the year depending on the desert station selected. In other words there may typically be around 150 to 270 days in the year in which rock temperature conditions are favourable to the salt hydration mechanism of rock decay occurring.

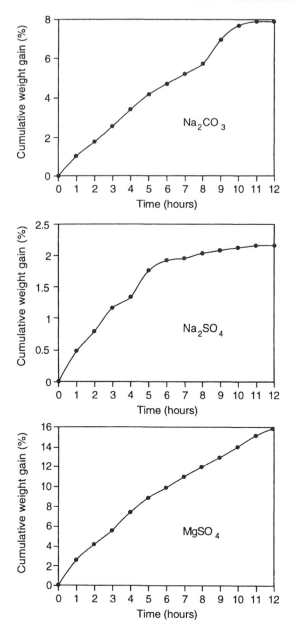

Figure 5.5. Uptake of moisture by three salt types under controlled temperature and humidity conditions. For explanation, see text

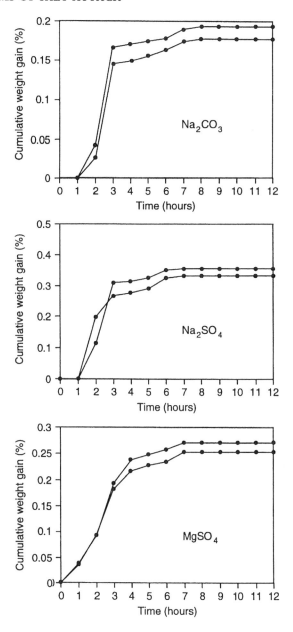

Figure 5.6. Uptake of moisture by blocks of York Stone in the presence of different salts. For explanation, see text

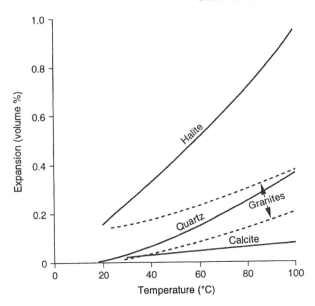

Figure 5.7. Differential thermal expansion of halite compared with quartz, calcite and granite. Modified after Cooke and Smalley (1968). Reproduced with permission

DIFFERENTIAL THERMAL EXPANSION OF SALTS

A further possible mechanism of rock disruption through salt action has been proposed by Cooke and Smalley (1968). They argue that the disruption of rock may take place because of the fact that certain salts have higher coefficients of expansion than the minerals of the rock in whose pores they occur. As Figure 5.7 illustrates, halite expands by around 0.9% between 0 and 100°C, whereas the volume expansion of quartz and granites is generally about one-third of that value. Gypsum and sodium nitrate are other common salts that have a relatively greater expansion potential than rock minerals (Table 5.8).

It is difficult to assess the actual importance of this process, and while some early experimental simulations (e.g. Goudie, 1974) did not suggest that it was very effective, much more work is required on this mechanism before its potential can be dismissed. Moreover, as Cooke et al (1993: 40) point out empirical studies of cliff weathering by Johannessen et al (1982) point strongly towards the effectiveness of this process.

They showed that, in the mid-latitude coastal environment of Oregon, the rate of cliff weathering was at least ten times faster on sunny south-facing cliffs than on shaded north-facing cliffs. This contrast is attributed to the expansion

Table 5.8. Coefficients of volume expansion between 20 and 100°C

Salts	
Halite	0.963
Gypsum	0.58
NaNO$_3$	1.076
Rock minerals	
Microcline	0.128
Orthoclase	0.049
Plagioclase	0.14
Quartz	0.36
Calcite	0.105
Olivine	0.21

Source: from data in Skinner (1966).

of sodium chloride on heating in the spray zone on the south-facing cliffs. The most important factor is probably the absolute range of temperature differences. Johannessen et al illustrated the effect of sodium chloride operating on quartz crystals (SiO_2) in a sandstone with a temperature change of 50°C. The basic equation of potential pressure they suggested is:

$$K_{NaCl}P + K_{SiO_2} = T(E_{NaCl} + E_{SiO_2})$$

where K is the linear compression a crystal undergoes when pressured, P is the pressure in kilobar, T is the change in temperature and E is the linear thermal expansion of a crystal during a heating episode (when $K_{NaCl} = 0.0014$/kbar, $K_{SiO_2} = 0.0009$/kbar, $E_{NaCl} = 0.000040$/°C, $E_{SiO_2} = 0.000005$/°C, and $T = 50$°C). In these circumstances: $0.0014P$/°C $+ 0.0009P$/°C $= 50$°C $(0.00004 + 0.000005)$ $(P = 0.97 \text{ kbar} = 970$ bar). The pressure of 970 bar (i.e. 970×10^5 Pa), they argued, is sufficient to break quartz crystals from their surrounding cement in the sandstone.

SLAKING

One possible role of the presence of salts is to cause accelerated slaking of clay-rich rocks (shales, mudstones, etc). The presence of high amounts of exchangeable sodium—normally expressed as the exchangeable sodium percentage (ESP)—has long been recognised as a cause of swelling and structure change in such materials (see, for example, Seedsman, 1986), particularly if they are smectite-rich. However, experimental studies of slaking rates for shales (Matsukura and Yatsu, 1985) did not indicate that the presence of salt accelerated the process of slaking. Indeed, it was found that the slaking rate of shale by salt water is slightly lower than that by distilled water. Nonetheless, extensive studies of mudrock and over consolidated clay formations in both the UK and the USA have shown that many of the rocks that are susceptible to breakdown have exceptionally high ESP rates (Taylor and Smith, 1986).

CORROSION EFFECTS ON REINFORCEMENTS

Many engineering structures are made of concrete containing iron reinforcements. The formation of the corrosion products of iron (i.e. rust) causes a volume expansion to occur. If we assume that the prime composition of such corrosion products is $Fe(OH)_3$, then the volume increase over the uncorroded iron can be four-fold. Thus when rust is formed on the iron reinforcements, pressure is exerted on the surrounding concrete. This may cause the concrete cover over the reinforcements to crack, which in turn allows the ingress of oxygen and moisture, which then aggravates the corrosion process. In due course, the spalling of concrete takes place, the reinforcements become progressively less strong, and the whole structure may suffer severe deterioration.

Rates of corrosion are accelerated in the presence of chlorides. Chloride ions may occur in a concrete because of the use of contaminated aggregates or because of penetration from a saline environment (Soroka, 1993: section 10.5). However, the electrochemical corrosion of metals can also be produced by sulphates (Hong Naifeng, 1994: 33) because there are often sulphate-reducing bacteria in a saline soil containing sulphates, which can cause strong corrosion of metals.

SULPHATE ATTACK ON CONCRETE

Sulphates can cause severe damage to, and even complete deterioration of, Portland cement concrete (Mehta, 1983; Bijen 1989). Although there is still controversy as to the exact mechanism of sulphate attack (Cabrera and Plowman, 1988), it is widely appreciated that the sulphates react with the alumina-bearing phases of the hydrated cement to give a high sulphate form of calcium aluminate known as ettringite ($3CaO \cdot Al_2O_3 \cdot 3CaSO_4 \cdot 32H_2O$).

Magnesium sulphate is particularly aggressive because in addition to reacting with the aluminate and calcium hydroxide as do the other sulphates, it decomposes the hydrated calcium silicates and, by continued action, also decomposes calcium sulphoaluminate (Addleson and Rice, 1991: 407).

The formation of ettringite involves an increase in the volume of the reacting solids, a pressure build up, expansion and, in the most severe cases, cracking and deterioration (Soroka, 1993: section 9.3.1). The volume change on the formation of ettringite is very large, and is even greater than that produced by the hydration of sodium sulphate.

Another mineral formed by sulphates coming into contact with cement is thaumasite ($CaSiO_3 \cdot CaCO_3 \cdot CaSO_4 \cdot 15H_2O$). This causes both expansion and softening of cement (Crammond, 1985) and has been seen as a cause of the disintegration of rendered brickwork and of concrete lining in tunnels (Lukas, 1975). A useful review of this mineral and its effects on Portland cement is provided by van Aardt and Visser (1975).

The chemical reactions involved in sulphate attack on compounds in hydrated cement have been summarised by Oberholster et al (1983). Sodium sulphate reacts as follows:

$$Ca(OH)_2 + Na_2SO_4 + 2H_2O \longrightarrow CaSO_4 \cdot 2H_2O + 2NaOH$$
$$4CaO \cdot Al_2O_3 \cdot xH_2O + 3CaSO_4 \cdot 2H_2O + H_2O \longrightarrow$$
$$3CaO \cdot Al_2O_3 \cdot 3CaSO_4 \cdot 32H_2O + Ca(OH)_2$$

or

$$3CaO \cdot Al_2O_3 \cdot 6H_2O + 3CaSO_4 \cdot 2H_2O + H_2O \longrightarrow$$
$$3CaO \cdot Al_2O_3 \cdot 3CaSO_4 \cdot 32H_2O$$

Magnesium sulphate reacts with hydrated cement compounds as follows:

$$Ca(OH)_2 + MgSO_4 + 2H_2O \longrightarrow CaSO_4 \cdot 2H_2O + Mg(OH)_2$$
$$3CaO \cdot Al_2O_3 \cdot 6H_2O + 3CaSO_4 \cdot 2H_2O + H_2O \longrightarrow$$
$$3CaO \cdot Al_2O_3 \cdot 3CaSO_4 \cdot 32H_2O$$

or

$$x \, CaO \, y \, SiO_2 + MgSO_4 + H_2O \longrightarrow CaSO_4 \cdot 2H_2O + Mg(OH)_2 + SiO_2$$

Various experiments have been undertaken to assess the decrease in strength that takes place when cements are immersed in different concentrations of these two salts (see, for example, Kayyali, 1989). Figure 5.8 show the results of an experiment undertaken by Kumar and Kameswara Rao (1994) using water with sulphate concentrations of 0, 100, 1000, 3000, 5000 and 10 000 ppm. The percentage loss in compressive strength at any one age was calculated with reference to the strength of concrete in freshwater at that age. The loss of compressive strength increases with salt concentrations and is high for all three types of sulphate used.

In a similar vein, Figure 5.9 shows the linear expansion that mortar prisms undergo when immersed in different sulphate solutions (Oberholster et al, 1983).

It is for this reason that concrete exposed to sulphate-bearing soils (e.g. gypsum crusts in deserts) or to sulphate-containing water should be made with sulphate-resisting cement or with a concrete with a high cement content (i.e. with a low water to cement ratio). High density and low permeability are important factors. The use of Portland–pozzolanic cements and pozzolanic admixtures is also recommended as an aid in the control of sulphate attack. As Table 5.9 shows, there are certain standards in Britain for concrete exposed to

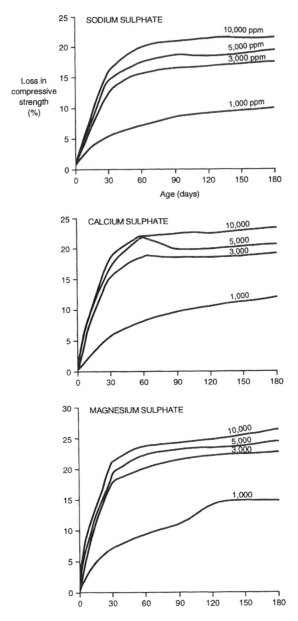

Figure 5.8. Loss in the compressive strength of concrete samples over 180 days in salt solutions of different strengths. Modified after Kumar and Kameswara Rao (1994: figures 3, 4 and 5) and reproduced by permission of Elsevier Science Ltd

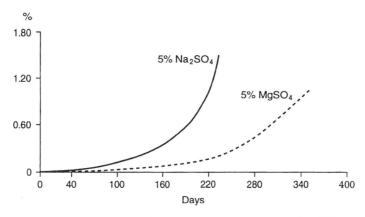

Figure 5.9. Linear expansion of mortar prisms immersed in different sulphate solutions. Modified from Oberholster et al (1983: figure 20, 400). Reproduced by permission of Elsevier Science Ltd

sulphate attack, while Figure 5.10 indicates the sulphate resistance of various types of cement.

SULPHATION AND CRUST FORMATION

The various components of the sulphation reaction are presented in Figure 5.11. Sulphur dioxide becomes oxidised (sometimes in the presence of catalysts) on moist surfaces to form sulphuric acid. The sulphuric acid then reacts with the stone in the following way:

$$CaCO_3 + H_2SO_4 + 2H_2O \rightarrow CaSO_4 \cdot 2H_2O + CO_2 + H_2O$$

According to Cooke and Gibbs (no date), where little moisture is present two alternative series of reactions may occur. Hydrated calcium sulphate may form first from the reaction of calcium carbonate and sulphur dioxide, as represented in the following equation:

$$CaCO_3 + SO_2 + 2H_2O \rightarrow CaSO_3 \cdot 2H_2O + CO_2$$

Subsequently, this hydrated calcium sulphate may become oxidised to form gypsum in the presence of catalysts. Alternatively, sulphur dioxide may react with bicarbonate solutions formed by a reaction with rainfall containing dissolved carbon dioxide as follows:

$$Ca(HCO_3)_2 + SO_2 \rightarrow CaSO_3 + 2CO_2 + 2H$$

The deposition of sulphates on material surfaces, often in the form of a gypsum crust (Figure 5.12), can be seen as both a consequence and a cause of

Table 5.9. Requirements for concrete exposed to sulphate attack (Note: recommendations for concrete in near-neutral groundwater only). ©Crown Copyright. Reproduced by permission of the Controller of HMSO

Class	Concentrations of sulphates expressed as SO_3			Type of cement		Requirements for dense fully compacted concrete made with aggregates meeting the requirements of BS 882 or 1047	
	In soil		In groundwater			Minimum cement content*	Maximum free water cement*
	Total SO_3 (%)	SO_3 in 2:1 water:soil extract (gl^{-1})	(gl^{-1})			$(kg\,m^{-3})$	ratio
1	Less than 0.2	Less than 1.0	Less than 0.3	Ordinary Portland cement (OPC) or rapid-hardening Portland cement (RHPC) or combinations of either cement with slag‡ or pfa§ Portland blastfurnace cement (PBFC)	Plain concrete†	275	0.65
					Reinforced concrete	300	0.60
2	0.2–0.5	1.0–1.9	0.3–1.2	OPC or RHPC or combinations of either cement with slag or pfa PBFC		330	0.50
				OPC or RHPC, combined with minimum 70% or maximum 90% slag‖		310	0.55
				OPC or RHPC, combined with minimum 25% or maximum 40% pfa‖			
				Sulphate-resisting Portland cement (SRPC)		280	0.55

3	0.5–1.0	1.9–3.1	1.2–2.5	OPC or RHPC, combined with minimum 70% or maximum 90% slag OPC or RHPC, combined with minimum 25% or maximum 40% pfa	380	0.45
				SRPC	330	0.50
4	1.0–2.0	3.1–5.6	2.5–5.0	SRPC	370	0.45
5	Over 2	Over 5.6	Over 5.0	SRPC + protective coating**	370	0.45

*Inclusive of content of pfa or slag. These cement contents relate to 20 mm nominal maximum size aggregate. To maintain the cement content of the mortar fraction at similar values, the minimum cement contents given should be increased by 50 kg/m⁻³ for 10 mm nominal maximum size aggregate and may be decreased by 40 kg/m⁻³ for 40 mm nominal maximum size aggregate.

†When using strip foundations and trench fill for low-rise buildings in Class 1 sulphate conditions, further relaxation to 220 kg/m⁻³ is permissible in the cement content for C20 grade concrete.

‡Ground granulated blastfurnace slag. A new BS is in preparation.

§Pulverized fuel ash to BS 3892: Part 1: 1982.

¶Per cent by weight of slag/cement mixture.

‖Per cent by weight of pfa/cement mixture.

**See BS 8102: 1990: Protection of structures against water from the ground.

Source: from Building Research Establishment (1986).

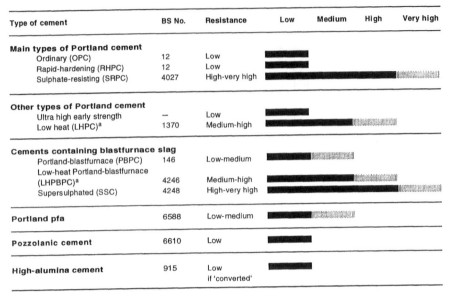

Type of cement	BS No.	Resistance	Low	Medium	High	Very high
Main types of Portland cement						
Ordinary (OPC)	12	Low				
Rapid-hardening (RHPC)	12	Low				
Sulphate-resisting (SRPC)	4027	High-very high				
Other types of Portland cement						
Ultra high early strength	–	Low				
Low heat (LHPC)[a]	1370	Medium-high				
Cements containing blastfurnace slag						
Portland-blastfurnace (PBPC)	146	Low-medium				
Low-heat Portland-blastfurnace (LHPBPC)[a]	4246	Medium-high				
Supersulphated (SSC)	4248	High-very high				
Portland pfa	6588	Low-medium				
Pozzolanic cement	6610	Low				
High-alumina cement	915	Low if 'converted'				

a Normally available in the UK to special order
b For general guidance only

Figure 5.10. Sulphate resistance of various types of cement (from *Building Research Establishment*, 1987). ©Crown Copyright. Reproduced by permission of the Controller of HMSO

weathering. This is especially true of limestones, where such a crust may be harder than the material on which it has developed. Other properties may also be different and may cause an acceleration in the speed of stone degradation. Amoroso and Fassina (1983: 264) identify three mechanisms that may account for this. Firstly, a variation in volume occurs because gypsum has a greater volume than the quantity of calcite it replaces. One volume of calcium carbonate forms over two volumes of hydrated calcium sulphate (Schaffer, 1932: 31). This causes expansive stresses in pores and cracks. Secondly, calcite and gypsum have different thermal expansion characteristics. The linear coefficient of the thermal expansion of gypsum is about five times that of calcite (Schaffer, 1932: 31). This difference may be further increased when blackened crusts develop because they tend to absorb a larger amount of radiation than white surfaces. Thirdly, the development of a crust will reduce the permeability of the material, which will in turn increase the water retention beneath and "all the ensuing adverse effects".

Sulphation can lead to blister development and lamination, possibly through a combination of the factors just described. This was the view of Schaffer (1932: 31) who argued

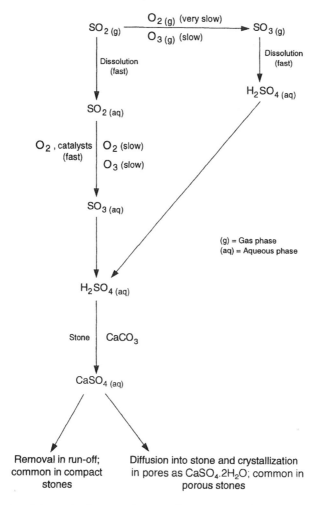

Figure 5.11. Schematic diagram showing the reactions of sulphur dioxide with moist stone

When the surface of a moist stone becomes warm by exposure to the sun or to exposure to a warm atmosphere, expansion occurs and, concurrently, evaporation takes place; crystals of calcium sulphate may thus be deposited in minute cracks produced by the expansion. Continued repetition of this process would tend to produce the observed expansion of the surface layers.

CHEMICAL WEATHERING BY ALKALINE AND OTHER SALTS

Some saline solutions can have increased pH levels. As Table 5.10 indicates, sodium carbonate is particularly alkaline and gives pH values in excess of 11.

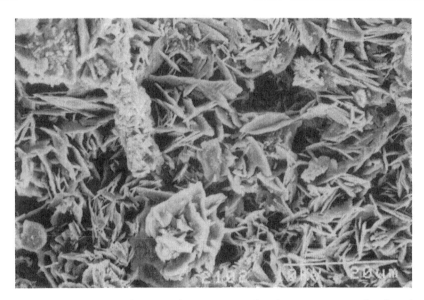

Figure 5.12. Scanning electron photomicrograph of gypsum crust developed on limestone. The scale bar at the bottom is 20 μm long (photograph by H. A. Viles)

Table 5.10. Characteristics of some laboratory salt solutions

Salt	Concentration (g l^{-1})	Conductivity (mS cm^{-1})	pH
NaCl	25	33.6	5.30
	50	60.4	5.20
	100	101.3	5.25
Na$_2$CO$_3$	25	24.4	11.25
	50	42.0	11.25
	100	63.7	11.20
Na$_2$SO$_4$	25	8.81	8.15
	50	38.7	6.30
	100	62.3	5.95
MgSO$_4$·7H$_2$0	25	8.81	8.15
	50	14.70	6.20
	100	24.10	6.35
NaNO$_3$	25	23.40	6.05
	50	40.80	5.65
	100	70.40	5.85
Sea water	25	31.1	8.10
	50	55.3	8.15
	100	93.4	7.60

Table 5.11. Reported pH values for selected lake waters

Lake	pH
Pyramid Lake, Nevada	9.4
Lake Hayq, Ethiopia	9.0
Van Gulu, Turkey	9.9
Corangamite Lakes, SE Australia	8.2–9.1
Lagoons, Tasmania	8.0–10.4
Lake Kivu, Africa	9.1–9.5
Lake Turkana, Kenya	9.5–9.7
Wadi Natrun, Egypt	10.9–11.0
Lake Shala, Ethiopia	9.9
Great Salt Lake, Utah	7.4–7.7
Deep Springs Lake, California	7.4–9.6
Harney Lake, Oregon	9.8
Abert Lake, Oregon	9.7

Source: miscellaneous sources in Goudie and Cooke (1984).

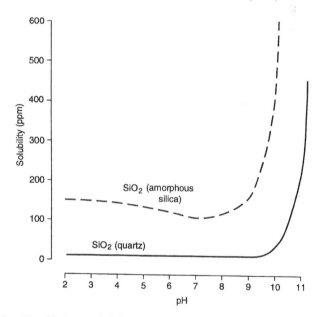

Figure 5.13. Equilibrium solubility (expressed in parts per million — ppm) of two forms of silica in water

Equally, as Table 5.11 shows, some saline lakes may also have pH values in excess of 9.

Why this is significant is that silica mobility tends to be greatly increased at pH values greater than 9. Indeed, according to various studies, the solubility of silica increases exponentially above pH 9 (Figure 5.13).

The presence of sodium chloride may also affect the degree and velocity of quartz solution. Experimental data on this are given by van Lier et al (1960), who found that at higher sodium chloride concentrations the solubility of quartz and the reaction velocity both increase. Sodium chloride can also accelerate the rate of limestone decay in the presence of gypsum. This was explained by Laurie and Milne (1926: 65)

> If roughly crushed crystals of calcite are placed in a dilute solution of sulphuric acid, there is a rapid effervescence for a few seconds followed by an almost complete cessation of action owing to the coating of calcium sulphate formed on the face of the crystals. If a solution of sodium chloride is now added, a vigorous action is at once set up and continues, doubtless due to the setting free of hydrochloric acid which, acting on the calcite, keeps the surface free from a continuous film of calcium sulphate.

Thus salts may contribute to rock weathering by leading to an acceleration of chemical weathering.

The growth of salt crystals may be able to cause the pressure solution of silicate grains in rocks, for the solubility of silica increases as pressure is applied to silicate grains. This is a mechanism that has been identified as important in areas where calcite crystals grow, as, for example, in areas of calcrete formation or petrocalcic horizon development (Goudie, 1973; Monger and Daugherty, 1991).

Schiavon et al (1995) have found petrographic evidence from granites in urban atmospheres which suggests that chemical reactions occur between granite minerals and weathering solutions which are responsible for the precipitation of gypsum. They found feldspar minerals that were partially or totally replaced by sulphate crystalline salts while still retaining their primary mineral outline and texture. What is the cause of this gypsum-induced pseudomorphism process? Schiavon et al (1995: 94) suggest the following:

> A likely mechanism for the gypsum replacement of silicate minerals may involve the localisation of silicate dissolution and gypsum precipitation to thin solution films between the silicate and sulphate phases; according to this model, the dissolution of silicates and the precipitation of gypsum should occur simultaneously and original textural features in the primary mineralogy can be preserved after replacement.

They go on to argue:

> ...the most important factors controlling the replacement mechanism are in fact degree of supersaturation of the gypsum precipitating solutions and relative crystallization pressures arising from these precipitation episodes...these pressures may then be responsible not only for displacive physical decay effects already reported on weathered granitic surfaces but also for the chemical replacement observed in this study (particularly when gypsum growth occurs within a spatially restricted area where no displacive movement is allowed such as inclusions within a rigid framework of a host mineral like quartz). The replacement is most effective on feldspars, probably because of their higher surface area susceptible to chemical attack (due to the abundance of

intracrystalline fractures and microcracks due to their higher chemical reactivity) together with their lower density with respect to quartz; mica's typical platy morphology and perfect basal cleavage does not seem to favour chemical decay but rather the physical detachment of the crystal's layers.

Other material components may also be mobilised by salt solutions. For example, when limestones are submerged in concentrated solutions, particularly of sodium carbonate, appreciable quantities of iron go into solution. The experimental study of Gillott (1979), discussed in Chapter 4, demonstrates the power of salt solutions to chemically etch limestones. Presumably in this case the pH of such solutions must be low. Work by Vergès-Belmin (1995) indicates that, in marble affected by sulphation in polluted air, pseudomorphism of calcite crystals to gypsum can occur. This suggests that, as in the granite situation reported by Schiavon et al (1995), gypsum precipitated into the stone can be dissolved to form sulphuric acid, which then attacks the calcite, before gypsum again becomes precipitated out of the solution.

In a similar vein, samples of stone from a wall affected by rising groundwater containing chlorides were studied by Kozlowski et al (1992b). The presence of both chloride and calcium ions brought about water adsorption when the relative humidity of air was greater than 20%, which is the pressure of water vapour in equilibrium with saturated calcium chloride solutions. This produces a strong rise in the moisture content of the stone at higher temperatures. At 100% relative humidity Kozlowski et al (1992b) found the moisture content of such stone to be 8 wt.%, which represents filling of about half the total water-accessible pore space of the sample. After the removal of the chlorides, the moisture content decreased to only 0.7 wt.%.

HYGROSCOPICITY, DELIQUESCENCE AND MOISTURE EFFECTS

Some salts are said to be hygroscopic and to have the ability to absorb water vapour from the atmosphere, and to yield a saturated solution. This phenomenon is also called deliquescence. In more precise terms a substance is said to be deliquescent when its water vapour pressure is lower than or equal to the water vapour pressure of the atmosphere. Under these conditions water vapour from the air will be absorbed by the deliquescent salt to form a solution. The water absorption will stop, and equilibrium will be reached, when the water vapour pressure of the saturated solution is equal to the vapour pressure in the atmosphere.

At a given temperature a particular substance will have a specific water vapour pressure (Piqué et al, 1992). Consequently the initiation of deliquescence depends on the value of the atmospheric water vapour pressure. Given that the relative humidity is a function of this parameter, it is possible to say that above a particular threshold value of relative humidity, called the equilibrium relative humidity for the deliquescent process (RH*), a substance

Table 5.12.   Equilibrium relative humidities of some salts from walls

Salt	0°C	5°C	10°C	15°C	20°C	25°C	30°C
$CaCl_2 \cdot 6H_2O$	41	37.7	33.7	—	30.8	28.6	22.4
$MgCl_2 \cdot 6H_2O$	33.7	33.6	33.5	33.3	33.1	32.8	32.4
$K_2CO_3 \cdot 2H_2O$	43.1	43.1	43.1	43.2	43.2	43.2	43.2
$Ca(NO_3)_2 \cdot 4H_2O$	59	59.6	56.5	54	53.6	50.5	46.8
$Mg(NO_3)_2 \cdot 6H_2O$	60.4	58.9	57.4	55.9	54.4	· 52.9	51.4
NH_4NO_3	—	—	—	—	—	61.8	—
$NaNO_3$	—	78.6	77.5	76.5	75.4	74.3	73.1
$NaCl$	75.5	75.7	75.7	75.6	75.5	75.3	75.1
Na_2SO_4	—	—	—	—	82	82.8	84.3
KCl	88.6	87.7	86.8	85.9	85.1	84.3	83.6
$MgSO_4 \cdot 7H_2O$	—	—	86.9	—	90.1	88.3	88
$Na_2CO_3 \cdot 10H_2O$	—	—	—	96.5	97.9	88.2	83.2
$Na_2SO_4 \cdot 10H_2O$	—	—	—	95.2	93.6	91.4	87.9
KNO_3	96.3	96.3	96	95.4	94.6	93.6	92.3
K_2SO_4	98.8	98.5	98.2	97.9	97.6	97.3	97

Source: Arnold and Zehnder (1990).

at a specific temperature will start to deliquesce. RH* at a known temperature can be calculated from the following equation:

$$RH^* = (P_{salt}/P_s) \cdot 100$$

where P_{salt} is the vapour pressure of the saturated solution of the salt and P_s is the atmospheric water vapour pressure at that known temperature. Table 5.12, derived from the work of Arnold and Zehnder (1990), shows the equilibrium relative humidities for a range of salts for a range of temperatures.

Halite is one salt that is hygroscopic and attracts moisture from the atmosphere, and Dutch experiments (cited by Winkler, 1981) show that a halite content of 4% in brick can attract nearly 10% of moisture into the masonry from the atmosphere at a relative humidity of 80%. Other salts are also hygroscopic, including calcium chloride and sodium nitrate.

The attraction of moisture into the pores of rocks or concrete can accelerate the operation of chemical weathering processes and of frost action (McInnis and Whiting, 1979) and the disruptive action of moisture trapped in rock capillaries is well known.

Hudec and Rigbey (1976) investigated the effects of sodium chloride on the water absorption of a suite of limestones and dolomites from North America. Their salt treatment consisted of immersing the rocks in a 3% by weight sodium chloride solution for a period of 72 hours. They found that at 98% relative humidity the salted rock absorbed about twice as much water as unsalted rocks.

Table 5.13. Moisture uptake in York Stone blocks in relation to temperature and humidity for concentrated solutions of NaCl and NaNO$_3$. Values given are net change in weight

Temperature (°C)	Humidity (%)								
	20	30	40	50	60	70	80	90	95
100% NaCl									
5	0	0	0	0	0	0	0	0.0421	0.1996
15	0	0	0	0	0	0	0.0313	0.1125	0.2060
25	0	0	0	0	0	0	0.0518	0.1138	0.3355
35	0	0	0	0	0	0	0.1667	0.3048	0.5692
45	0	0	0	0	0	0.0127	0.2723	0.5568	0.7702
100% NaNO$_3$									
5	0	0	0	0	0	0	0	0.0051	0.0510
15	0	0	0	0	0	0	0.0118	0.0461	0.0708
25	0	0	0	0	0	0	0.0235	0.0493	0.0767
35	0	0	0	0	0	0	0.0937	0.2109	0.3461
45	0	0	0	0	0	0.0086	0.1206	0.3243	0.4119

Table 5.13 shows the results of another simple study of this phenomenon using two salts: sodium nitrate and sodium chloride. Blocks of York Stone, a fine-grained sandstone, with dimensions of 3.5 × 4 × 5 cm were soaked in a saturated solution of the salts for 24 hours and then dried to a constant weight at 50°C and 50% relative humidity. Using an environmental cabinet they were subjected to a range of temperatures between 5 and 45°C. Their weights were recorded as they were subjected to increasing humidities between 20 and 95%. It is evident that at relative humities above around 70–80% the blocks took up moisture from the atmosphere, leading to a weight gain of 0.77% for sodium chloride and 0.41% for sodium nitrate. The same process was undertaken using lower concentrations of salt (Figure 5.14) and there is a clear increase in water uptake with salt concentration (and thus salt content in the rock pores).

The presence of hygroscopic salts may be especially important in polar regions and this has been noted in the context of Antarctica by Campbell and Claridge (1987: 140)

> Where salt concentrations are high and water is present, as in the many undrained basins and saline lakes of Antarctica, the soils may be damp, or even saturated with water, which, because of the high salt concentration, does not freeze, even though soil temperatures may be relatively low.

They also remark

> The presence of water-soluble salts within the soil is virtually essential to chemical weathering processes, which cannot proceed in the absence of moisture. Although very little moisture is available in Antarctica, concentrated salt

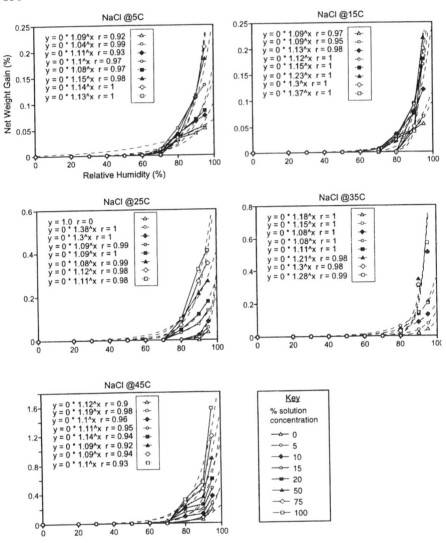

Figure 5.14. Uptake of moisture by rock at different temperatures and humidities for different concentrations of sodium chloride (see text for explanation)

solutions may remain unfrozen and, thus, chemical weathering and movement of ions may occur within the crevices of rocks and the surfaces of apparently dry rocks, and may lead to mineral transformations.

One of the earliest discussions of the importance of sodium chloride deliquescence was that by Wallace (1916: 32), who noticed the effects of brine springs that issued from the Devonian limestones and dolomites at the foot of the Manitoba escarpment in Canada

The salt-flats where the springs reach the surface are devoid of vegetation and studded with ice-carried boulders. These are representative of the pre-Cambrian igneous series of North-Central Canada — granites, gneisses, and epidiorites. They have suffered intense chemical disintegration, large boulders having been reduced to half their original size. Different minerals have been affected to different extents, but not even quartz or garnet has escaped corrosion. Ferromagnesians have been most intensely affected; and gneissose structures, hardly noticeable on unweathered surfaces, stand clearly revealed. The striking difference between the action of these brines and that of sea-water calls for explanation.

Thin crusts of salt gather, during the summer months, on the flats and around the boulders. The salt is somewhat deliquescent; and thin films of brine are drawn, by surface tension, over the surface of the boulders. Water in contact with the atmosphere is a powerful disintegrant. Alkalies are removed as chlorides or carbonates, and silica and alumina are precipitated as gels, separately or in combination. The gels exercise selective adsorption on the salts of the brine, alkali being taken up and the brine being left richer in the acid radicals. The brine is thereby rendered a more active disintegrating agent, and the process goes on continuously. The function of the dissolved salts is considered to be twofold: (1) they provide a thin film of liquid in contact with atmospheric oxygen; (2) owing to partial absorption by colloids, they provide an acid residual solution, which is a powerful corrosive agent.

SALT AND BIOLOGICAL WEATHERING

In recent years organisms have been increasingly recognised as effective agents of weathering in a diverse range of situations. As we demonstrated in Chapter 3, a range of organisms grow in even the most salt-affected, hostile environments such as sabkhas and soot-encrusted gypsum crusts. As yet, however, there has been little consideration over how salt and biological weathering processes might interact — apart from the recognition that many micro-organisms can precipitate sulphates and oxalates, and thus may be active in creating salts which then cause damage to rocks and building materials. Organisms themselves are capable of effectively weathering surfaces through a combination of physical and chemical means. Thus organisms such as lichens have a tremendous water-holding capacity and volume changes on wetting and drying can be considerable. Where such lichens are intimately bound to a rock surface (with part or all of their thallus growing within the rock) this swelling and shrinking can exert considerable destructive pressures.

In terms of chemical effects, many organisms exude acids and other chemicals which are capable of etching vulnerable minerals, such as calcium carbonate. The experimental study of Koestler et al (1985) showed how fungi artificially inoculated onto calcite and dolomite substrates under laboratory conditions could produce serious etching within a matter of days on crystal faces. Such effects will increase the surface area of minerals available for further weathering, and may increase the ease of ingress of water into rocks. In other cases, biochemical processes associated with organisms can lead to mineral transformations, producing new minerals which may be more susceptible to subsequent weathering. Finally, organisms can exert an indirect control on weathering on material surfaces. Where a biofilm or thicker cover of organisms exists on a surface it may have significant control over the environment, perhaps reducing the effective porosity through blanketing the surface, or altering the colour and thermal response characteristics, or protecting the surface from airborne pollutants.

Based on these types of organic weathering influences, we can propose three possible modes in which biological and salt weathering may interact: mutually exclusive; mutually reinforcing; or largely independent. In the mutually exclusive mode, salt and biological weathering can be seen as competing risks which do not occur under the same conditions. This seems to be the case in areas of the central Namib desert, where salt-affected surfaces are too unstable and rapidly decomposing to support lichens, and lichen-covered surfaces are found away from the major salt-affected areas. On many buildings gypsum crusts, salt efflorescences and lichen and biofilm growths are spatially segregated, each being favoured by different microenvironmental conditions. Alternatively, biological and salt weathering might have a synergistic, mutually reinforcing relationship where, for example, micro-organisms provide an initial etching of a surface, which then opens the way for salt attack. Conversely, salt attack could roughen a surface sufficiently for organisms to colonise. The biological production of salts would also come into the category of a synergistic relationship. Finally, a largely independent relationship between the two sets of processes could be envisaged, where, for example, salt impregnates and affects the stone at depth, with a biofilm growing on, and influencing weathering at, the surface.

There is an urgent need for well-designed experiments to provide further information on the reality of salt and biological weathering interactions. Although a lot of field evidence has also been collected from a diverse range of environmental settings, it is often difficult to interpret processes from field samples.

SALT AND FROST

One controversial issue in weathering studies is whether the presence of salts accelerates the rate at which frost action operates, and, if it does, why this should be. As we have seen in Chapter 4, some laboratory studies (see, for example, Goudie 1974 and Williams and Robinson, 1981) have indicated that some rocks disintegrate more rapidly when they are frozen after soaking in salt solution rather than in pure water, and various studies have shown that de-icing salts can promote the breakdown of concrete (see, for example, Litvan, 1972) by freeze–thaw. However, in some laboratory simulations (see McGreevy, 1982) salts may reduce or even prevent frost weathering.

This whole issue is important in terms of understanding weathering in high latitude coastal situations (e.g. on shore platforms) (see, for example, Trenhaile and Mercan, 1984), to understand the effects of de-icing salts on road surfaces and in engineering structures such as bridges, and also because salts generated by acid rain could conceivably enhance frost action.

Williams and Robinson (1991), in a useful review, have looked at some of the mechanisms that might explain why under some conditions salts could accelerate frost weathering.

1. Surface sealing. Salts that have accumulated on the outer layers of a rock surface as a result of surface evaporation could block pores and seal the surface. Water beneath such a sealed surface might be less able to escape by extrusion during freezing, thus increasing the stresses created within the rock.

2. Combined growth of salt and ice crystals. The combined action of salt crystal growth and frost action might be greater than they would be individually. Moreover, as salt crystals form they might tend to reduce the amount of pore space available for ice formation.

3. Osmotic pressure. The growth of ice crystals tends to expel solutes included in water and they diffuse into the unfrozen portions of the rock creating osmotic pressures which might in rocks with a particular micropore structure be sufficient to cause breakage.

4. Expansion of ordered water. If frost damage to sorption-sensitive rocks results from the expansion and contraction of absorbed, ordered water on clay surfaces, then because the presence of salts delays freezing there will be a greater amount of ordered water that can undergo expansion and, as the temperature falls, there is consequently more damage in the presence of saline solutions.

5. Greater saturation. Because of the hygroscopic nature of some salts, rocks containing salts can take up more moisture from the atmosphere than rocks that do not contain saline solutions. The presence of greater amounts of moisture in the rock pores could make them more susceptible to frost weathering.

6. Increased water mobility. Low concentrations of dissolved salts might increase the rates at which water moves to ice crystals on freezing. This could facilitate crystal growth and increase the crystallisation pressure.

7. Leapfrogging of the freezing front. The rejection of salts by growing ice crystals may cause them to accumulate beyond the advancing freezing front, thereby depressing the freezing point. In the zone where salts become concentrated freezing may be temporarily prevented so that the freezing front may jump this zone, reforming on the other side. An unfrozen layer might be expected to experience pressure as the adjacent frozen layers expand.

8. Reduced rate of freezing. If the freezing of a dilute salt solution is slower than that of a pure water solution, then the resulting ice crystals might be larger than usual and thus more effective at causing rock disintegration.

9. Corrosion. The corrosional effects of salts might, especially in carbonate rocks, act in a multiplicative manner with frost to produce rock disintegration.

These hypotheses largely lack experimental verification and no effort has been made here to judge their relative significance but, as Williams and Robinson (1991: 352) conclude

> The search for a single theory of frost and salt weathering would seem doomed to failure. The mechanisms will vary from site to site according to the freezing regime, the salts present and their concentration, the rock type and possibly other factors.

The relative importance of different salts in this context was looked at experimentally by Jerwood et al (1990a, 1990b). They found that magnesium sulphate actually inhibited the breakdown of cubes subjected to intense freezing, that sodium sulphate greatly enhanced rock breakdown under both mild and intense freezing regimes, while sodium chloride increased breakdown only under intense freezing conditions. The concentrations of salts are also highly important in affecting the rates of breakdown, though the links are highly complex. For example, Trenhaile and Rudakas (1981) found in experiments designed to simulate the weathering of shore platforms that the greatest distintegration tended to be associated with sea water solutions of about half normal salinity.

CONCLUSIONS

In this chapter we have presented a range of information on the mechanisms of salt weathering, drawn from a combination of field observations, laboratory experiments, and theoretical models. It is clear that the chemistry involved is hugely complicated and we are still a long way from a complete explanation of how salts act under different conditions. However, an increasing number of studies seems to be illustrating that the processes conventionally divided into chemical and physical forms of salt attack are more realistically thought of as often acting together. Furthermore, salt action is clearly allied to other weathering processes under real environmental conditions.

6 Geomorphological Implications of Salt Weathering

INTRODUCTION

Geomorphologists have made a substantial contribution to the study of the salt weathering hazard through their concern with the processes, rates of change, material resistance and the identification of those landscapes where salt plays a major part. Such an interest in salt attack is not surprising as salt weathering has been implicated in the development of a wide range of different geomorphological phenomena over a diversity of scales. At one extreme it has been seen as a cause of splitting of individual grains of sand, while at the other it has been regarded as a potential factor in the development of major erosion surfaces (Figure 6.1). Because of the distribution of salts and the environmental conditions necessary for such salts to be effective agents of weathering (as discussed in Chapter 3), salt weathering is of particular geomorphological importance in three main areas: arid, coastal and urban environments. However, as with weathering processes in general, its overall contribution to landscape evolution has not been fully assessed, although increasing information on the rate of salt attack (as presented later in this chapter) provides a more solid basis for such an assessment. As we will see, it has proved difficult to interpret the exact roles of salt action in the development of geomorphological features, largely because of the difficulties of inferring process from form. By this we mean that it is often hard to find evidence of what has caused certain landforms by looking at the landforms themselves, even in great detail.

The role of salt in promoting large landforms is difficult to assess, but the possibility that salt weathering can cause widespread "haloplanation" or "slope equiplanation" has from time to time been mooted. In the driest portion of the Atacama Desert, for example, "the completely barren landscape has a smoother lunar appearance" (Abrahams and Parsons, 1994: 29), and Abele (1983) asserted that these smoother slopes were restricted to the zone of coastal fog below an altitude of about 1100 m, where salt weathering was intensified by almost daily wetting and drying in the fog zone. He held that normal dissection of the landscape prevailed above this elevation. The central Namib Desert inland from Swakopmund and Luderitz also displays large expanses of smoother plains with rock that appears to have

Figure 6.1. Limestone residuals in eastern Abu Dhabi, United Arab Emirates. They rise above the general level of the salty sabkha surface and appear to be the result of the progressive wasting of the limestone as a result of salt attack (photograph by A. S. Goudie)

been deeply shattered by salts and here also salt weathering in the zone of frequent fogs has been seen as a major formative factor (Lageat, 1994; Beaudet and Michel, 1978) (Figure 6.2).

When we consider smaller features the role of salt is more testable, and in this chapter we concentrate on five main aspects: the development of closed depressions; the formation of various types of cavernous phenomenon; the weathering of shore platforms; the production of fine particles; and the formation of gypsum crusts and associated blistering and sealing.

These five aspects do not, however, cover the role of salt weathering as a geomorphological process. It has also been proposed to play a part in some other phenomena, including tor formation (Wellman and Wilson, 1965; Watts, 1981), rock labyrinth development (Selby and Wilson, 1971), the formation of pedestal rocks (Figure 6.3) (Chapman, 1980) and the development of micro-weathering forms of a polygonal type (Miotke and Hodenberg, 1980). Whether it plays a part in the formation of inselbergs was the subject of a minor discussion between Bruckner (1966) and Wellman and Wilson (1965). It also probably plays a major part in the form of certain desert surfaces, including stone pavements. This is partly because of its ability to cause ground heave (Horta, 1985) and displacive growth (Searl and Rankin, 1993), but also because of its role in particle size reduction and gravel shattering (Amit et al, 1993). Furthermore, the weathering of surface

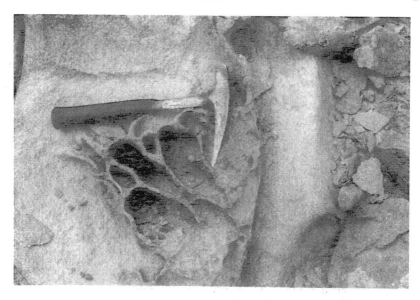

Figure 6.2. Alveoles developed in granite at Luderitz, Namibia. These forms appear to be developed especially well in close proximity to the Atlantic Ocean coast, where salt loadings in the atmosphere are high. Frequent fogs cause frequent wetting of the rock surface and this hastens the salt weathering process in this environment (photograph by A. S. Goudie)

Figure 6.3. The wick effect and the upward migration of saline solutions on the margins of the Death Valley Playa, California, USA have contributed to the formation of this pedestal rock, which is about 1.5 m high (photograph by A. S. Goudie)

materials modifies water movements in the upper portion of the soil and thus can modify the nature of erosive processes (Martin-Penela, 1994). Salt weathering may also prevent or retard the formation of desert varnish (Dragovich, 1994) and help to explain differences in the rates of weathering between continental and maritime environments (André, 1995: 124).

SALT WEATHERING AND PAN FORMATION

Pans or closed desert depressions (see Goudie and Wells, 1995 for a review) are a widespread landform type and salt weathering has long been recognised as a potential mechanism for their development and enlargement.

Some of the best known but most controversial desert depressions occur in the Western Desert of Egypt (see Al-Izz, 1971 for a review). Some of them are very large and in some cases their bases lie below sea level (e.g. Fayum, 45 m; Khargha, 18 m; and Qattara, 143 m). They have been the subject of debate for a considerable time. Ball (1927) believed that they were essentially deflational in origin, but others pointed out the position of the depressions with respect to important geological boundaries and proposed that the primary factor in their formation was tectonics (e.g. Gindy and El Askary, 1969), a view that was doubted by Said (1962). More recently there has been discussion about the role of fluvial incision in the Miocene and karstic disruption of drainage (see, for example, Albritton et al, 1990). Certainly the great thickness of the limestone caprock in the vicinity of Qattara, and the depression's shear volume $(20\,000\,km^3)$ have posed problems for any simple deflational hypothesis. This is what has made the tectonic and fluvial hypotheses attractive. Gindy and El Askary (1969: 622) wrote: "All available evidence indicates tectonism as the initial cause of the formation of the Siwa depression; wind deflation was not even a secondary agent". They went on (p. 624): "the initial surface depression that resulted from the tectonic activity captured the surface drainage of the wet post-middle Miocene climate. Water erosion sculpted the ancient landform at Siwa and enlarged the initial tectonic depression to nearly its present size. Wind deflation and exudation, contrary to general belief, were minor factors in the evolution of the depression". Albritton et al (1990), working on Qattara, supported the fluvial hypothesis but largely rejected the tectonic factor (see also Gindy, 1991), and as a working hypothesis proposed (p. 952) "that the Qattara Depression originated as a stream valley that was subsequently dismembered by karstic processes during the Late Miocene epoch and afterward was deepened by deflation and otherwise modified by mass wasting and fluviatile processes". They believed that its evolution had to be seen in terms both of the history of palaeo-drainages of the Nile and of sea level changes in the Mediterranean basin.

The role of salt weathering in preparing material for subsequent deflation is recognised as a possibility by Gindy and El Askari (1969: 623), who observe that fragments and large blocks of limestone and chalk readily disintegrate and disappear when they fall into adjoining swampy saline water (as in the case of the fallen columns of the Ptolemaic temple at Agourmi). They believe that the salt could be derived from Miocene saliferous shales. Haynes (1982: 104) has championed the combined power of salt weathering and deflation in creating the depressions of the Western Desert:

> Whereas some major depressions appear to coincide with faults (for example, the Kharga Depression), the negative topographic areas are principally due to deflation of shales or sandstones weakened by salt weathering following desiccation of former lakes, or by evaporation of an exposed capillary fringe ... The process can be witnessed today at the remote salt lake of Merga in Northwestern Sudan. Nubian sandstone around the western margin of the present lake is very friable due to the high content of efflorescent salts ...

He continues (p. 106):

> From observations in many depressions of the Western Desert, from the Qattara in the north to Merga in the south, I am of the opinion that most if not all are the result of eolian corrosion and deflation of beds weakened by leaching of cement and salt efflorescence ... The mechanism is self-enhancing, once started by the development of an initial area of internal drainage. Each wetting allows new salts to form upon drying, and crystallization and recrystallization of expanding salts weakens the cementing of the sedimentary matrix and loosens grains for their plucking by the wind ... Once the water table is approached by the basin floor, efflorescence by evaporation from the capillary fringe would further aid deflation as long as evaporation exceeded recharge.

The rift valley lakes of East Africa also show signs of salt attack on their floors and sides. The lakes are largely characterised by brines which contain a substantial component of sodium carbonate. Wherever bedrock exposures have been examined in close proximity to such lakes, clear signs of weathering have been identified. At Lake Stefanie (Chew Bahir) in southern Ethiopia a transition can be witnessed as one moves towards the *playa* floor from well rounded volcanic masses, to mushroom-shaped pedestals, to planed off structures. By Lake Magadi in Kenya veins and blisters of trona (sodium carbonate) appear to be disrupting lavas, and stone-floored tracks across the playa show signs of substantial decay of their surfaces as a result of aggregate disintegration. Exposed floors around the Galla lakes to the south of Addis Ababa show a whole array of microweathering forms, including *gnammas*. However, trona is not the only salt involved. At Magadi, for example, Smith and McAlister (1986) found that trachyte lavas were suffering from flaking and granular disintegration caused by two salts: halite and thermonatrite. They also describe extensive tafoni development and honeycomb formation.

Likewise, salt weathering has been implicated in the development and pattern of distribution of pans in South Africa. Pans tend to be especially prevalent on non-resistant sedimentary rocks from the Dwyka and Ecca.

One of the primary reasons for the concentration of pans on some lithologies appears to be that these units rapidly disintegrate when exposed to salt weathering. Du Toit (1906: 257) observed the effect of salt on bedrock at Groot Chwaing Pan in the northern Cape: "The brine solution attacks the flagstones, and they become covered with peculiar pitting and hollows . . . the flags being highly micaceous and fissile, break up readily, and in this way by disintegration and wind erosion, the pan is gradually being deepened".

The preferential development of pans on rocks such as the Ecca shales may be accounted for in part by their susceptibility and in part by the fact that they may also be a major source of salt. Thus Bruiyn (1971: 123) has related the retarded development of pans on the Beaufort sandstones compared with the Ecca and Dwyka shales to the observation that "salts are relatively scarce in Beaufort sediments due to their original depositional environment in shallow fresh-water swamps". Weathering of the Dwyka and Ecca shales appears to release considerable quantities of salt, and Hugo (1974) has noted the relationship between this salt liberation and pan development: "Most of the pans occur on Dwyka and Ecca Shales, and the brines are characterised by the presence of relatively much sodium sulphate in comparison with other impurities, mostly calcium and magnesium sulphates and chlorides".

Another major pan area where salt weathering has been seen to play a formative part is in Western Australia (Figure 6.4). Woodward (1897: 365)

Figure 6.4. A small salt pan inland from Perth, Western Australia. It was in situations such as these that Jutson and others proposed that a combination of salt weathering and aeolian deflation contributed to the formation and enlargement of salt pans (photograph by A. S. Goudie)

recognised that salt accumulation leads to rock disintegration and prepares materials so that they are in a suitable form for deflational removal. He talked of "salt wind-planed flats":

> The salts deposited upon these surfaces consist very largely of gypsum, which upon crystallizing out in the clay causes it to split up, thus allowing the weather to act upon it, which quickly converts it into an almost impalpable powder, which the wind distributes far and wide over the surface of the country.

This was a view that was also adopted by Jutson (see, for example, Jutson, 1918) to account for the development of over-steepened cliffs ("breakaways") and flat ("billiard-table") rock floors. He also believed that salt weathering helped to detach small particles and that they might also be lifted up by crystal growth so that they would be exposed to wind removal (Jutson, 1950: 254)

> The writer believes that on the floors and at the foot of the cliffs of the salt lakes, disintegration of rocks takes place owing to the crystallisation of salts from solutions which are brought to the surface by capillarity. On the floors fairly soft rocks easily crumble, and at the foot of the cliffs the action is best observable where the rocks are still fairly undecomposed. There the action appears to cause a rapid flaking of the rock which results in an undermining of the cliff at its base.

Jutson provides an impressive plate of a greenstone cliff being thus undermined at Yelladine Road in the Yilgarn Goldfield. More recently, Clarke (1994) has drawn attention to the role of halite efflorescences on the north and west shores of lakes in the same area in causing cliff sapping in felsic and sedimentary rocks.

There is some field evidence that salt weathering can proceed quickly on pan floors. For example, dramatic evidence of salt attack can be seen at various pans in South Africa, where bricks, concrete structures and stone walls on pan surfaces are constantly being renewed in response to their breakdown in the presence of sodium sulphate and sodium chloride (Goudie and Thomas, 1985: 17). Similarly, a limited amount of monitoring of the rate at which rock and cement blocks takes place has been attempted, most notably in Tunisia (Goudie and Watson, 1984). Close to Zarzis, in the southern part of that country, there is a salty pan or playa called the Sebkret el Mebabeal. It is an erstwhile arm of the sea that during the Holocene has become isolated by a combination of tectonic subsidence and barrier growth. As with other basins of the paralic type, halite (sodium chloride) is the main type of precipitate. Goudie and Watson placed blocks of York Stone and concrete in a transect across the playa surface and then recovered them after a period of six years. During that period the concrete blocks severely broke up, so that of 36 original blocks used in the traverse, just one was still identifiable. Thirteen of the original York Stone blocks remained, and these were mostly split into a series of parallel "bread-and-butter" slithers (Figure 6.5). York Stone is a hard Lower Carboniferous siliceous sandstone which is much used

Figure 6.5. In a simple monitoring experiment in southern Tunisia, sandstone blocks were emplaced (A) and suffered severe weathering (B) in a halite-rich environment after just six years (photograph by A. S. Goudie)

for paving in the UK on account of its supposed durability. This simple monitoring experiment, which is being repeated elsewhere, indicates the rapidity with which clasts on a salty playa surface may disintegrate and also, incidentally, provides a demonstration of the power of sodium chloride attack.

TAFONI, ALVEOLES, HONEYCOMBS AND OVERHANGS

The world's rock surfaces are frequently pitted with a range of cavernous weathering forms of different sizes. Such forms occur throughout the world from polar valleys to tropical coastlines. The larger varieties tend to be known as tafoni (Figure 6.6) whereas smaller cavities are more generally known as alveoles or honeycombs (Figure 6.7). Tafoni may be particularly widespread in arid areas and they are recorded from a number of the world's deserts. Alveoles and tafoni are also common in coastal environments, forming on many cliffs. They also occur in a wide range of rock types (Table 6.1). A comparable table on honeycombs is provided by Mustoe (1982, table 1). Kirchner (1996) argues that only rocks with relatively closely spaced discontinuities (e.g. jointing, foliation, bedding planes) such as shales, slates or suites of strongly interbedded sedimentary rocks are unaffected by cavernous weathering forms.

Salt weathering has often been seen as associated with the development of cavernous weathering forms (e.g. Martini, 1978), but as Smith and McAlister (1986: 446) have rightly pointed out:

> there has been a tendency in field studies of salt weathering for either dangerous circular argument or a form of "guilt by association". In the first case, the presence of certain landforms in an area is used to infer the activity of salt weathering mechanisms, without incontrovertible evidence that the landforms

Figure 6.6. One landform feature that may well be associated with salt weathering is the formation of cavernous weathering forms called tafoni. This example is developed in limestone in Ras Al Khaimah (photograph by A. S. Goudie)

Figure 6.7. Alveoles forming in the Hawkesbury Sandstone outcrops in Sydney, New South Wales, Australia. Many examples of alveoles have been attributed to salt weathering (photograph by A. S. Goudie)

derive solely from these mechanisms. In particular, tafoni, which are commonly attributed to salt weathering, may . . . be convergent forms produced by different mechanisms, of which salt weathering is only one possibility. Whereas, in the second case it is assumed that because certain salts are abundant in an area, and because rocks are being weathered, then those salts must be responsible for the weathering.

There is yet another problem. Even if salt weathering can be proved to be the cause of rock disintegration, we still need to establish how it is that cavernous forms develop through this mechanism. We also need to establish whether salt's role is essentially mechanical or chemical.

Mottershead and Pye (1994) see tafoni formation as a two-stage process. The first stage involves case-hardening of the rocks (in their case, greenschist). This is achieved by chemical weathering of ferromagnesian-rich silicate minerals which releases solutions that are transported along joints. Some of it is reprecipitated, causing the greenschist to be compartmentalised into patina-encased rock masses. The second stage involves breaking this patina and the exposure of the subjacent greenschist to a saline environment (as revealed by SEM observations and microprobe analyses). However, in their example the role of the saline environment was to cause the dissolution of minerals along crystal boundaries, leading to chemically induced granular disintegration. Haloclasty itself was rejected as the mechanism. They suggested that the minerals of which the greenschist is composed (amphiboles, chlorite and

Table 6.1. Examples of tafoni

Location	Rock types	Reference
SW USA	Igneous, sandstone, sandy shale, conglomerates	Blackwelder (1929)
Grand Canaria	Phonolitic lavas	This book
Luderitz, Namibia	Granites	This book
Ras Al Khaimah	Limestone	This book
Utah, USA	Calcareous sandstone	Mustoe (1983)
Baffin Island, Canada	Granites and migmatites	Watts (1979)
South Australia	Granite	Bradley et al (1978)
Taylor Valley, Antarctica	Granite and gneiss	Prebble (1967)
South Devon, UK	Greenschists	Mottershead and Pye (1994)
New South Wales, Australia	Hawkesbury Sandstone	Young and Young (1992)
Tenerife	Volcanic tuffs, etc.	Höllermann (1975)
Magadi, Kenya	Trachyte lavas	Smith and McAlister (1986)
South Australia	Granite	Dragovich (1969)
Aravalli Hills, India	Granite	This book
Tuscany and Elba, Italy	Granodiorite, metamorphosed conglomerates	Martini (1978)
Montserrat, Catalonia, Spain	Conglomerates	This book
Atacama, Chile	Granitoid rocks	Segerstrom and Henriquez (1964)
Bahrain	Dolomitic limestones	Doornkamp et al (1980)
SW USA	Rhyolites, granodiorites, etc.	Kirchner (1996)
Algerian Sahara	Palaeozoic sandstones and granites	Klaer (1993)
Petra, Jordan	Ram Sandstone	This book

epidote), by possessing mineral structures in which the silicate components are held together only by metallic cations, might be particularly susceptible in a marine salt environment to weathering by ion-exchange processes.

Similar arguments were advanced by Pye and Mottershead (1995) to account for honeycombs developed on sea walls composed of Carboniferous sandstone in the west of England. They believe that the weathering was assisted by the presence of salt which caused the enhanced swelling and contraction of clay minerals within the rock.

Another champion of the chemical role of salt has been Young (1987), who studied cavernous weathering forms developed in sandstones in coastal regions of Western Australia and New South Wales. Scanning electron microscopy revealed that although salt was present, so was the intense etching of quartz grains. She argued that sodium chloride increased the rate of quartz dissolution and concluded (p. 566)

Soluble salts, particularly sodium chloride, in water percolating through quartzose sandstones play an important role in promoting the granular disintegration of the interiors of weathering caverns. That role does not appear to be the physical forcing apart of grains by crystallization–hydration pressures but, rather, the enhancement of the rate of silica solution and thence widening or creation of voids in the sandstone.

The suggestion that the chemical role of salt could be important in the development of cavernous weathering forms has also been made elsewhere. For example, McGreevy (1985) looked at honeycomb formation in a Carboniferous sandstone in a coastal environment in Northern Ireland. He identified the presence of gypsum and suggested that while SEM observations indicated that the crystallisation of salt in pores could easily dislodge quartz grains to promote the granular disintegration of the sandstone, the etching of the quartz grain surfaces also attested to chemical weathering activity within the rock.

Mustoe (1982) studied how honeycombs developed on coastal cliffs bordering Puget Sound in the north-west of the USA. He, in contrast, found no evidence of mineral alteration and so believed that the prime process was the crystallisation of salts derived from sea spray.

Bradley et al (1978) undertook a study of the salts associated with granitic tafoni in South Australia. They found that total salt content (comprised primarily of halite and gypsum) was two to thirteen times more abundant in tafoni flakes than in massive granite and accepted the association between salts and flaking granite as a genetic one. They believed that the saline material was derived from fluid inclusions within the granite and that this material dissolves when moisture seeps into the rock, to be precipitated as halite and gypsum when evaporation pulls the fluid back to the surface and renders it supersaturated. They undertook a study of the salt content of different portions of a boulder with tafoni development (Figure 6.8) and remarked (p. 653): "The sheltered upper parts of tafoni are most favourable for the retention of moisture, and this is where salts are concentrated and flaking most active. Salts either do not move to the more exposed rock surfaces, due to inadequate moisture, or they are flushed away from such locations by rainwater". However, the amount of salt involved in the process was low — only 0.4% of salt was present in the tafoni flakes.

Höllermann (1975) studied the tafoni of Tenerife in the Canary Islands. He argued that they were associated with case hardening, therefore lending support to the later observations of Mottershead and Pye (1994). He also analysed the salts involved in the tafoni and found that the primary salts were sodium chloride (dominant), sodium sulphate and calcium sulphate.

Another attempt to quantify the presence of salt associated with tafoni was made by Mustoe (1983). Working in the Capitol Reef Desert of Utah, USA, he found that efflorescences were present and used atomic absorption spectrometry and X-ray diffraction to identify which ions and salts were present. Among the salts he found were gypsum, hexahydrate, bloedite and natron.

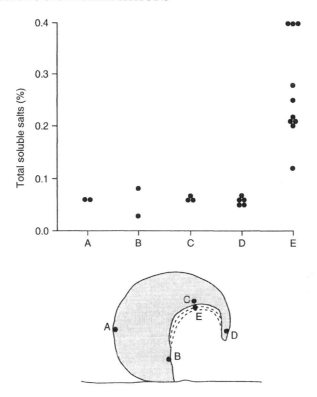

Figure 6.8. Total soluble salts versus sample locations for an Australian tafoni. (A) Outside of boulder; (B) solid (non-flaking tafoni wall); (C) massive rock behind flakes; (D) visor; and (E) tafoni flakes. Modified from Bradley et al (1978: figure 4) and reproduced with permission

Like Höllermann, he argued that salt crystallisation was the most important cause of sandstone disintegration within the tafoni.

One particularly original approach to the role of salt crystallisation in tafoni development was made by Rögner (1986), who worked on cavernous forms in the Negev Desert. He monitored temperatures within a tafoni and found anomalous variations which he attributed to the process of salt crystallisation out of highly concentrated solutions. He also recorded which salts were present behind scales and flakes. They included halite, gypsum, epsomite, hexahydrite and sylvite. He believed that the presence of these salts and the fact that temperature observations showed that crystallisation occurred within the tafoni were both reasons for believing that "salt weathering is the main process in flaking, scaling and origin of the tafoni in Machtesh Hagdol" (p. 1271).

It may be that once formed tafoni develop through a positive feedback effect. This has been expressed thus by Smith and McAlister (1986: 456)

. . . once a hollow is initiated it creates an environment in which weathering is favoured, weathering in turn extends the hollow to produce an optimum form in which weathering is further enhanced, and so on. Such re-enforcement cannot continue indefinitely and a condition must eventually be reached where the weathering rate is reduced. This could, occur, for example, where a cavern becomes so deep that it either prevents the ingress of moist air, or rock temperature variations are reduced to the point where precipitation and evaporation no longer occur.

They argue that on exposed cliff faces salts would be deposited by the outward migration of salts derived from within the rock, but that they would be removed by subsequent rainwash, whereas salts precipitated in hollows would be protected from such leaching and so could cause salt weathering to occur, which would progressively expand the hollow.

Tafoni development has been noted on the undersides of leaning gravestones in eastern England (Osmaston, 1995 pers. comm.), where the migration of salts from groundwater has been implicated. Vertical stones are not affected, and Osmaston offers the following explanation, which also involves the role of rainwash:

Rain falling on the upper side of a leaning headstone or driven by wind against either side of an exposed vertical stone will wash the surface free from salt, the flow being largest and most effective at the base of the stone where the effects of any capillary rise from the soil would otherwise be concentrated. It will then form a high-volume flow into the soil on this side of the stone, flushing salts from it. On the sheltered side of the stone the reverse will occur.

Smith and McAlister (1986) suggest that case hardening is not a prerequisite for tafoni development, and in this they followed Dragovich (1969: 171) in her study of South Australian examples

. . . the walls of tafoni are frequently composed of fresh rock. Examination of the rock immediately behind the overhang revealed no signs of "bleaching" which could represent a zone of mineral depletion, nor had weathering activity been more successful behind the overhang than on other parts of the tafoni roof.

The existence of an overhang on all tafoni would be expected if all outcrop surfaces were case-hardened. However, recently initiated cavernous surfaces at Taratap and elsewhere lack this characteristic feature. It is therefore postulated that the overhang forms as a result of the rapid weathering of the tafoni back wall, and that there is no necessity to invoke the presence of a more resistant outer layer of rock. The development of case-hardening on north-facing (exposed) surfaces has been proposed, but, if this is so, the direction of tafoni openings does not reflect it.

Case-hardening may be present on some outcrops in which tafoni have formed, but there is no evidence to suggest that tafoni initiation is in any way controlled by the presence or absence of such surfaces. The origin of tafoni must therefore be more closely related to factors influencing the rate and intensity of weathering attack at particular points on the rock surface.

Dragovich proposed that the weathering processes involved in tafoni formation were stress development accompanying the chemical alteration of minerals and the expansion and contraction of clays situated between minerals. She reported that salt crystals were "present in discernible quantities in only a few coastal tafoni and in no inland hollows, although positive results were obtained from tests for the presence of traces of chlorine and sulphide in fragments detached from tafoni walls" (p. 179). However, while she was prepared to admit that salt weathering might play a part, she argued that "the conditions necessary for Salzsprengung are those which will cause disaggregation whether salts are present or not".

The question of the role of case-hardening was also addressed by Winkler (1979). He believed it was the primary cause of tafoni development, a view which was challenged by Bradley et al (1978) and by Conca and Rossman (1985). Conca and Rossman argued that it was not so much exterior case-hardening, but interior decay and kaolinite formation that was crucial.

Although both the case-hardening model and the positive feedback model account for the development of a hollow behind a visor or small orifice, we also need to consider the precise form of the hollow, because many tafoni show signs not only of inward growth, but also upward growth. This may be partly because downward growth may be restricted by the fall of spall onto the tafoni floor, but it may also be related to the receipt of warmth from the sun. Segerstrom and Henriquez (1964), following Blackwelder (1929), suggest that dampness is a key feature of hollows and promotes weathering. Given, it is argued, that the lower part of a cavity is more likely to be dried out by sunshine than the more shaded upper part, there will therefore be a tendency for the cavity to grow inward and upward. The flow vectors of moisture within boulders may also be a powerful control of tafoni form (Conca and Rossman, 1985; Conca and Astor, 1987).

For a cavity to form it is necessary for there to be a mechanism to remove flakes and spalls. Wind may play a major part, but organisms such as pack rats, lizards and beetles may also contribute (Blackwelder, 1929). Although early workers thought that the actual excavation of a cavity might be achieved by wind abrasion, many tafoni occur in environments where sand blasting does not occur or they have an aspect (i.e. the leeward side of a boulder) that precludes such a mechanism.

There is very little hard and fast information about the speed at which tafoni development may occur. Several experiments indicate that it can be a rapid process. The first of these was undertaken in Bahrain by R. U. Cooke and A. S. Goudie (see Doornkamp et al, 1980). They sprayed a 33 cm wide band with red paint on a cavernous overhang, developed in a siltstone member of the Al Buhayr Carbonate Formation and possessing a slightly damp encrustation of salts, which included gypsum. When the site was revisited after an interval of five months, approximately 75% of the paint had been removed, largely in the form of small flakes and silt ("rock flour") which had accumulated on the cavern floor.

Another experiment was undertaken in Japan by Matsukura and Matsuoka (1991). They looked at the degree of tafoni development on the faces of marine cliffs on three uplifted shore platforms with different altitudes and with known ages of emergence. They found that the rate of deepening of tafoni is an exponential function of time, with the highest rate (1.67 mm per year) at the initial stage. They averred (p. 56) that "If salt weathering is essential for tafoni development, this decline with increasing depth may reflect the inner wall of the tafoni becoming more difficult to desiccate due to decreasing exposure to sun and wind".

What becomes evident from this and other work is that salt weathering is not invariably the cause of tafoni formation, but that there is strong evidence that salts are often present in tafoni backwalls and flakes and that the salts cause disintegration both through physical and chemical mechanisms. The cavernous form may result either from the breaching of a case-hardened exterior or from positive feedback effects leading to the enlargement of an initially small hollow. In coastal locations the salts may be derived from spray, and the same may apply in the vicinity of desert lake basins (Butler and Mount, 1986), but elsewhere the salts may be derived from within the rock mass by seepage. A model of tafoni development is shown in Figure 6.9.

A final class of weathering-related feature that needs consideration is the formation of amphitheatrical valley heads. Such features result in part at least from a process called "groundwater sapping", and recent research has shown that seepage processes are important in producing them in areas such as the Kalahari of southern Africa and the sandstone terrains of the Colorado Plateau. In arid regions where seepage occurs we also have salt exudation, so that salt weathering may contribute to the undercutting processes that causes valley head recession (Figure 6.10). This is certainly what has been suggested by Laity and Malin (1985: 206–207) in the context of the Colorado Plateau.

Ground-water sapping occurs at the heads of canyon and locally along sidewalls as a result of seepage that emerges from the Navajo Sandstone, just above the contact with the Kayenta Formation, and from perched ground water isolated by lithologic discontinuities in the Navajo Sandstone. The sites of groundwater seepage from cliffs are commonly marked by cavities and alcoves that range in size from centimetres to tens of metres. The rock surfaces within a cavity are soft and crumbly and may be scaly in appearance. Salt crystals often occur as thin (100–200 μm in thickness), discontinuous encrustations on the surfaces of the back wall and roofs. Beneath the salts, there is a layer of sandy calcareous material . . . The wedging action of the calcite decreases the strength of the Navajo Sandstone by separating interlocking sand grains.

The important sapping role of salt weathering was also something that struck Wellman and Wilson (1965: 1058)

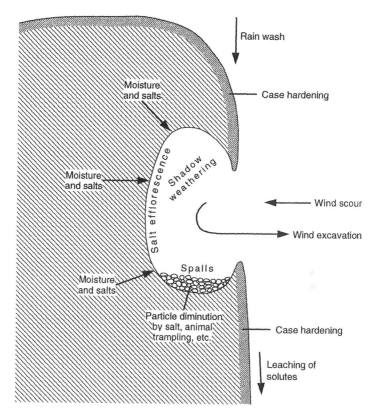

Figure 6.9. A model of tafoni development

In its attack, salt weathering is similar to most other erosive processes in having a well-defined base level below which it cannot act, but it differs from most other erosive processes by acting most rapidly on the lower and not on the upper side of rock surfaces. It is thus a powerful undercutting agent that constantly tends to steepen slopes to the limit of rock strength and is responsible for many unusual topographic forms such as cavernous weathering, coastal and desert platforms, some kinds of tors, and at least some hills that have been described as inselbergs.

Certainly, extreme cavernous weathering can cause cliffs to be oversteepened (Figure 6.11), and Hume (1925: 214) believed that on the Eocene Ma'aza Limestone Plateau in Upper Egypt this had progressed sufficiently to cause them to become "absolutely unscaleable". In the Asama volcano region of Japan, notch formation at the base of pumice flow cliffs has also been attributed to salt fretting (Matsukura and Kanai, 1988).

Figure 6.10. An actively weathering cliff face in the Namib Desert in the Khan Canyon. Seepage of saline water is causing oversteepening to occur and this mega-tafoni to develop (photograph by A. S. Goudie)

COASTAL WEATHERING AND SHORE PLATFORMS

Because of spray effects and the deposition of saline aerosols in proximity to coastlines, salt weathering may well be an important process on coastlines. The shore zone is one that is subjected to the alternation of phases of salt water inundation and partial or total drying out. Shore platforms and the lower parts of cliff profiles are exposed to a range of wetting and drying regimes, depending on their location relative to mean tide level (e.g. supratidal, intertidal or subtidal) and the local environmental conditions which may favour salt spray, splash or washing. Thus landforms such as shore platforms may be subjected to significant salt attack (Figure 6.12). A general review of this possibility is provided by Trenhaile (1987). Among the

Figure 6.11. Salt weathering by gypsum and halite have contributed to the development of this large overhang in a dolomite cliff that bounds a small closed depression excavated by wind. In the deep shade of the overhang, the presence of moisture enables *schattenverwitterung* (shadow weathering) to occur. Large amounts of rock flour are thereby produced, which are then susceptible to wind deflation. The location is Bahrain, Arabian Gulf (photograph by A. S. Goudie)

phenomena to which he refers are disaggregated rocks, a number of microtopographical forms and "water-layer levelling".

Dunn (1915) and Coleman et al (1966), working in Australia, have discussed the formation of disaggregated rocks on high tidal flats in the Northern Territory and in Queensland, respectively. A combination of sea spray, thorough wetting by high tides, and high rates of evaporation provide ideal conditions for salt crystallisation to occur. Likewise in the seasonal tropics of West Africa, Tricart (1962) has stressed the importance of coastal salt weathering on coastlines with long dry seasons or high levels of insolation. Mediterranean environments may fulfil these criteria (Moses and Smith, 1994) as well as the seasonal tropics. However, coastal salt weathering forms have also been identified as being important in the moist temperate environment of south-west England (Mottershead, 1982), and in Oregon (Johannessen et al, 1982). Nevertheless, Guilcher and Bodéré (1975), after a study of corrosion forms on volcanic rocks in a range of different climatic environments bordering the Atlantic Ocean, suggest that the efficacy of coastal salt weathering declines as the latitude increases.

One of the most broad-ranging attempts to explain coastal landforms in terms of salt weathering processes was that undertaken in Oregon by

Figure 6.12. A shore platform developed in volcanic rocks at Cape Schank on the Mornington Peninsula, Victoria, Australia. The platform surrounds a stack and is developed on both the exposed and landward sides of the feature. Salt weathering has been implicated in the formation of such landform assemblages (photograph by A. S. Goudie)

Johannessen et al (1982). They maintained that salt weathering was implicated in the formation of alveolar surfaces, marine stacks and in promoting aspect-related asymmetrical cliff retreat. They found that sunny, south-facing, hard-rock cliff retreat was many times more rapid than the shady-side cliff retreat whether subjected to waves or protected from wave attack, and they attributed this to the drying and thermal expansion of salts in the sunny situations. They also pointed to the protective role of freshwater streams (p. 29). In areas where streams flowed out of small valleys and discharged over the sea cliff as waterfalls, angular points of rock in line with the stream channel projected seaward beyond the general line of the cliff escarpment. The washing by freshwater prevented salt attack and so reduced the rate of cliff retreat in proximity to the stream compared with elsewhere along the cliffs.

There have been many studies on the weathering and erosion processes and the development of distinctive microtopographies found on limestone coasts, and salt has been seen to be possibly implicated in a range of ways. Characteristically, the evolution of features such as the coastal notch and suites of coastal karren forms has been ascribed to a combination of chemical denudation (through the dissolution of calcium carbonate in seawater), mechanical erosion (though abrasion) and biological erosion (through the action of a whole range of micro- and macro-borers and grazing animals). In recent years many studies have shown conclusively the efficacy of biological erosion in such settings (e.g. Schneider, 1976; Trudgill, 1985; Kelletat, 1988), as well as a host of consolidating and protective activities of other organisms (e.g. Focke, 1978, Dalongeville, 1995), although in other areas abrasion (e.g. the limestone coasts of peninsular Malaysia studied by Tjia, 1985) and chemical erosion (e.g. carbonate-cemented beachrock and aeolianites along the east coast of South Africa studied by Miller and Mason, 1994) have been given a prime role. Salt action seems to have been relatively neglected in general, although its potential role in some parts of the coastal profile has been identified by some workers, as well as its influence on water chemistry. Moses and Smith (1994), studying a supra-tidal zone comprised of a cliff and upper platform in southern Mallorca, suggest that salt weathering dominates the splash and spray zones here, with algae and lichen action becoming dominant at higher levels. Trudgill (1985), studying limestone shore platform bioerosion in County Clare, Eire, suggested that there was circumstantial evidence for salt weathering occurring in the upper lichen zone of the shore platform, but that the chemical action of lichens was also apparent. Most other studies have ignored the potential haloclastic role of salt within limestone coastal profiles, except in the production of tafoni and alveoles.

There has long been a debate about the potential of seawater to dissolve limestone, as most seawater is supersaturated with calcium carbonate and thus unable to dissolve more. However, detailed investigations such as those of Miller and Mason (1994) indicate that, especially in tidal pools emersed for reasonably long periods, the pH of water may decline and it once more becomes aggressive to calcium carbonate. Similarly, the water in such pools may become more concentrated through evaporation, allowing some salts to precipitate out. Further work is needed on the processes at work in such tidal pools, and the role of salt and other processes in their evolution.

PRODUCTION OF FINE PARTICLES

One of the most important consequences of all weathering processes is to cause a diminution in the size of particles making up a rock, sediment or soil. Salt weathering is not unique in this way and there is now a substantial body of information that indicates that it is effective in producing silt-sized material — the "rock flour" of field observers. Such material may be susceptible to translocation by wind and/or water and so may contribute to the erosion of rock surfaces and to the subsequent deposition of material such as loess.

Indeed, the origin of loess was one problem which stimulated early research into the particle size diminution produced by salt weathering. In particular, people have been concerned to answer the question "Can loess have a desert as opposed to a glacial origin?" It is an issue that still intrigues scientists (Smalley, 1995).

The literature on loess contains several suggestions of mechanisms that might produce silt-sized (generally in the range 20–50 μm) quartz particles from larger fragments. Those most frequently favoured include glacial grinding (Smalley, 1966), insolation weathering, frost action (Zeuner, 1949) and spalling during wind transport (Smalley and Vita-Finzi, 1968). However, in an analysis of the world distribution of loess, Smalley and Vita-Finzi (1968) concluded that most deposits can reasonably be explained in terms of a glacial grinding origin, followed by deflation from glacial, fluvio-glacial and fluvial environments, and ultimate deposition in periglacial areas. They argued that few loess deposits either occur in desert areas or are derived from them (the loess deposits of the Negev being a significant exception). Despite numerous assertions that some loess is of "hot desert" origin, Smalley and Vita-Finzi (1968) argued that the only indigenous process in hot deserts likely to produce silt-sized particles is that of spalling during aeolian transport by the grain-to-grain impact of relatively angular sand particles. They considered that this process was probably less efficient per unit time than glacial grinding (although its inefficiency might be offset by continuous production from the large areas of sand available). Keunen (1969) went further, arguing that the process does not produce new quartz particles between 20 and 50 μm in mass.

However, although there are relatively few well-reported loess deposits within deserts, they do exist, as in Iran (Fookes and Knill, 1969), Iraq (Kukal and Saadallah, 1970), Tunisia (Brunnacker, 1973), the Negev (Yaalon, 1965) and on Bahrain (Brunsden et al, 1979). Indeed Smalley and Krinsley (1978) recognised this fact. In addition, and despite Smalley and Vita-Finzi's arguments, it remains possible that some loess deposits marginal to deserts could have been derived from them, because studies of dust storms and of deep sea cores show that substantial quantities of silt-sized material are removed with great frequency from deserts (Goudie, 1978). Dust storms are also capable of long distance transport.

Pedro (1957a, 1957b) undertook a granulometric determination on the material liberated by the disintegration process under laboratory conditions and found that substantial quantities of silt were liberated and that different salts produced different size distributions of debris. For example, more material finer than 20 μm was produced by calcium salts, whereas magnesium salts tended to produce more debris in the 20–200 μm range.

Goudie et al (1979) designed an experiment to test the hyphothesis that silt-sized quartz particles, the main component of loess, might be produced by the salt weathering of aeolian sand. They used an environmental cabinet to undertake the simulation, adopting a temperature and humidity cycle recorded at Wadi Digla, near Cairo, in August 1922. This was selected as being reasonably typical of the ground-surface climate in an extremely hot, dry environment. They used dune sand from southern Africa and analysed the

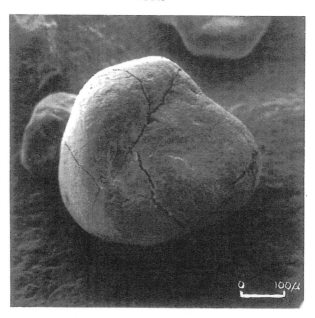

Figure 6.13. An experimental simulation by Goudie et al (1979) demonstrated that sodium sulphate could cause the disintegration of individual grains of aeolian sand. The salt has succeeded in causing fractures to develop. These in turn lead to the development of silt-sized particles, which may provide a source for desert loess

products of weathering with a Quantimet image analyser and a scanning electron microscope. The salt employed was sodium sulphate. The treatments produced cracking of sand-sized quartz grains and the production of appreciable silt-sized debris (Figure 6.13).

Subsequently, other experimental simulations have been undertaken. Pye and Sperling (1983) subjected dune sand and regolith material to sodium sulphate attack, also using the Wadi Digla cycle. They used SEM analysis and a Coulter counter to look at the debris liberated in the simulation. Like Goudie et al (1979) they succeeded in producing silt-sized material and they also made some comments on the surface textures they observed (Pye and Sperling, 1983: 60):

> The shapes and surface textures of grains affected by salt weathering are dominated by angular breakage surfaces which are indistinguishable from those produced by frost action and crushing. It is therefore unlikely that surface features seen under the SEM can be used to identify a salt-weathering origin for fossil sediments. Such features clearly are not, as some earlier workers maintained (e.g. Smalley & Cabrera, 1970), diagnostic of a glacial grinding origin.

Fahey (1985) took the argument two steps further by subjecting actual rocks (dolomite, shale, sandstone and schist) to simulated weathering and by using cold rather than hot climatic conditions. Samples subjected to mild solutions

Figure 6.14. An illustration of the wick effect and the disruptive consequences of upward salt migration on the margins of the Great Salt Lake, Utah, USA. The gypsum has migrated upwards in solution into the telegraph pole (photograph by A. S. Goudie)

of either sodium sulphate or magnesium sulphate showed greater breakdown than those subjected solely to water and also produced substantial quantities of fines.

Goudie (1986) simulated what he termed "the wick effect" (Figure 6.14) and used a laser granulometer to characterise the grain-size characteristics of debris liberated from York Stone — a Lower Carboniferous sandstone. Once again large amounts of silt-sized material were formed. Goudie and Viles (1995) used a Negev cycle (daily ground surface temperature range of 10–41°C and relative humidity between 100 and 30%) to test the breakdown of not only York Stone, but also of five different limestones. As Table 6.2 shows, all samples generated some material in the clay size range (i.e. less than 2 μm) with the mean for all 17 samples being 10.66%. The mean amount of

Table 6.2. Grain size characteristics of liberated debris (expressed as percentage at each grain size)

Sample	Salt type	$<2\mu m$	$2–63\mu m$	Silt and clay	$63–355\mu m$	$>355\mu m$
Bath Stone						
NB2 20*	Na_2SO_4	23.34	40.01	63.35	3.34	33.31
NB2 63*	Na_2SO_4	11.28	30.00	41.28	20.80	37.36
NB1 20	Na_2CO_3	19.90	40.32	60.22	3.58	36.20
NB1 63	Na_2CO_3	17.08	41.66	58.14	4.76	36.50
Ketton Stone						
NK2 20	Na_2SO_4	3.56	18.71	22.27	10.05	68.68
NK2 63	Na_2SO_4	2.92	16.95	19.81	9.08	71.05
NK1 20	Na_2CO_3	5.73	16.66	22.39	1.38	76.23
NK1 63	Na_2CO_3	4.02	15.44	19.96	6.49	74.05
Portland Stone						
NP2 20	Na_2SO_4	11.94	44.58	56.54	17.66	25.82
NP2 63	Na_2SO_4	18.51	51.91	70.92	19.86	9.72
NP1 20	Na_2CO_3	12.53	48.09	60.62	20.20	19.18
Portuguese Stone						
NMC2 20	Na_2SO_4	9.27	58.15	61.42	9.19	23.39
NMC2 63	Na_2SO_4	9.62	48.69	54.31	9.42	32.37
NMC1 63	Na_2CO_3	6.28	46.70	52.18	0.21	46.81
York Stone						
NY2 63	Na_2SO_4	8.50	85.90	94.40	5.60	0.00
NY1 63	Na_2CO_3	2.54	41.22	43.16	50.36	5.88
Chalk						
NC2 20	Na_2SO_4	16.78	65.24	82.02	8.21	9.77
Mean		10.66	42.27	53.13	11.26	35.65

*20=After 20 cycles; 63=after 63 cycles.

material produced in the silt size range (2–63 μm) was 42–47%, giving a total mean fines content of 53.13%. The only rock type that produced relatively low amounts of fines was Ketton stone. This consistently produced a large amount of debris that was coarser than 355 μm (mean of four samples is 52.50%). The explanation for this is that the Ketton stone is largely composed of coarse ooids of large diameter, which were detached more or less intact from their matrix of micrite cement. By contrast, the York Stone, a fine-grained sandstone, produced very little coarse debris.

One of the interesting findings of the grain size analysis is the bimodality or trimodality that some of the samples display. This is especially true for three of the oolitic Jurassic limestones (Bath, Portland and Portuguese). They have a high clay-sized fraction, another peak in the medium silt-sized fraction, a trough in the fine and medium sand-sized fractions, and another peak coarser than 355 μm. The first two of these peaks may reflect, respectively, the breakdown of the micritic matrix (micrite is generally finer than 2–4 μm) and the release of ooids and other particles inherent in the original petrographic structure of the rocks. The third modal fraction may reflect aggregates of ooids that have not been fully detached from their cementing matrix, larger fossil fragments, and similar material.

The way in which material properties influence the nature of liberated debris was also shown by Smith et al (1987). They subjected a Carboniferous sandstone to simulated hot desert conditions in the presence of sodium chloride, sodium sulphate and magnesium sulphate and examined the debris with a scanning electron microscope. They reported (p. 199)

> Micrographs and particle size analysis show that most debris consists of more or less intact sand-sized grains liberated from the parent rock. However, coarse and medium silt-sized material, often characterised by fresh fracture surfaces, was also observed. Examination of the parent rock suggests that this silt-sized material originates from two sources: the breaking away of silica cement and coatings of secondary silica from around sand-sized grains, and the micro-fracturing of quartz sand grains. Microfracture patterns observed in thin sections appear to result from point-loading by adjacent grains, which results in compressive and/or shear stresses sufficient to fracture individual grains.

The picture that emerges from these laboratory simulations on a wide range of rock types is that salt weathering can indeed produce quantities of silt-sized material and may therefore provide a source of material for incorporation in loess. However, there is a second method available to pursue this issue, and this is to look at the grain size characteristics of material weathered in the field in saline environments.

One context where this has been done is on arid zone alluvial pans. Beaumont (1968) reported that pebbles transported onto the salty margins of the Great Kavir in Iran appeared to disintegrate rapidly and this was a phenomenon that Goudie and Day (1980) sought to quantify in the case of fan sediments in Death Valley, California (Figure 6.15). They found on one of their fan traverses down to the salt playa that there were large spreads of pebbles existing on the salt playa fringe which, though plainly fractured by the weathering process (Figure 6.16), had still not fallen apart. They were kept together, to a large extent, by the cementing action of the salt. These pebbles were collected with care in bulk. Back in the laboratory they were immersed in water, which caused them to break down rapidly as the binding salt was dissolved. The grain size characteristics of the disintegrated particles (derived from a range of igneous and volcanic rocks) were determined, and showed that significant amounts of silt and clay (averaging 8.7% and ranging from 1.6 to 27.8% for eight boulders) were formed by the weathering process.

We have also carried out a comparable study on valley bottom pebbles in a stream in the central Namib Desert, where a salty spring leads to salt precipitation in the stream bed (Figure 6.17). Two gneiss pebbles and one granite pebble were leached in water. The silt and clay percentages of the disintegrated material were 12.85, 13.23 and 25.57%. Two other weathered

Figure 6.15. A Landsat image of Death Valley, California, USA. The white salt efflorescences on the distal ends of alluvial fans are clearly evident. Ground observations show that in such locations the clasts making up the fans suffer from rapid and extreme weathering

Figure 6.16. Disintegrated clasts on an alluvial fan as it enters the saline zone at Badwater, Death Valley, California, USA. Disintegration appears to be rapid and to produce large amounts of fine-grained material (photograph by A. S. Goudie)

Figure 6.17. Salt wedging taking place along joints in gneissic rocks in a stream bed in the Namib Desert near Rossing, Namibia (photograph by A. S. Goudie)

samples of gneiss from other salty environments in the Namib gave values of 20.59 and 73.48%. The average value for all five samples is 29.14% silt and clay. Of this the clay proportion is small—averaging only 1.65% (Table 6.3).

One more recent study in a coastal environment in southern England was undertaken by Mottershead and Pye (1994). They collected granular debris released from cavernous weathering features derived from greenschist. Their tafoni floor samples had an average silt content of 61% and their tafoni wall samples of 59%. Negligible proportions of material finer than medium silt were present, suggesting to them that this grade represents the effective lower limit of this weathering process. This agrees with the findings of the Namib study.

Table 6.3. Grain-size characteristics of rocks weathered in saline environments in the Central Namib Desert

Sample No.	Percentage 2–63 μm	Percentage <2 μm	Total silt+clay (%)
94 N 32 A (gneiss)	12.49	0.36	12.85
94 N 32 B (gneiss)	12.70	0.53	13.23
94 N 32 C (granite)	23.49	2.08	25.57
94 N 14 (gneiss)	19.12	1.47	20.59
94 N 28e (gneiss)	69.67	3.81	73.48
Mean	27.49	1.65	29.14

GYPSUM CRUSTS AND ASSOCIATED BLISTERING AND SPALLING

Although reported mainly from buildings, gypsum encrustations have been found on some near-urban natural rock outcrops (e.g. on carbonate cliffs of the Calanque, near Marseille, southern France studied by Del Monte and Sabbioni, 1984), but whatever their settings, such crusts are features of geomorphological interest as well as being of concern to stone conservators, as discussed in Chapter 7. As discussed earlier, there are two main types of, usually blackened, crusts which occur widely on building stone surfaces. Firstly, there are the "thin black layers" of Nord and Ericsson (1993), which form widely on non-carbonate-rich stone and are a deposit usually rich in iron and often depleted in gypsum. Secondly, there are crusts formed at least partly as a reaction product on and in carbonate-rich stone, which are rich in gypsum. It is this second type which is of most geomorphological interest as it can produce a thick build-up on the stone surface which ultimately starts to blister and peel away, causing unsightly damage.

There has been much debate over how such crusts form. Some workers view the process as a simple growth from the original stone surface outwards, while others see a more complex process involving outward growth coupled with inward growth and the replacement of original calcite crystals. Camuffo et al (1983) recognised three distinctive zones on urban limestone façades: white zones frequently washed by rainwater and covered with a thin layer of reprecipitated calcite; grey zones sheltered from rainwater which are chemically unaltered and covered with a thin layer of dust; and, finally, black areas soaked by rain, but not subjected to runoff, where gypsum and carbonaceous particle-enriched crusts develop. A slightly different model, specific to Portland limestone, has been proposed by Duffy and Perry (1996), which recognises white, washed areas, blackened encrustations and delaminating areas (where gypsum and reprecipitated calcite block near-surface pores, producing blistering and delamination). The studies of Camuffo et al (1983) suggested that the gypsum layers grow from the stone–gypsum interface outwards with the older, outer gypsum crystals eventually breaking off.

The more porous the stone, the more likely it is that sulphate-rich solutions can percolate into the stone, producing gypsum formation at depth within the stone, as well as near the surface (as found by McGee and Mossotti, 1992, who recorded greater gypsum accumulation on test briquettes of limestone than similar ones made of marble). As the studies of Vergès-Belmin (1995) and Lal Gauri et al (1989) have shown, gypsum crusts can be complex entities, with identifiable layers present where different associations of processes occur. An outer, blackened crust is found over a thinner layer where gypsum has replaced calcite crystals.

Blistering may occur simply as a result of gypsum crust growth, where volume increases and thermal conductivity differences between the crust and underlying stone cause the crust to begin to spall away. However, the situation may be more complex as soluble salts may accumulate beneath the crust, aiding blister development through crystallisation processes (Whalley et al, 1992). Certainly, studies of the friable zone within and beneath blistering crusts have revealed high contents of a range of salts. Studies of series of photographs (Viles, 1993a and b) have revealed that blisters grow irregularly over time, and may continue to develop even after air pollution has declined, suggesting that other processes may be contributing to their development.

HOW FAST DOES SALT WEATHERING OPERATE ON NATURAL ROCK OUTCROPS?

The speed with which salt can cause the disintegration of buildings of known age suggests that under suitable conditions it may operate at such a rate as to lead to the rapid weathering of natural rock outcrops. The same implication arises from the speed with which some laboratory simulations cause rock disintegration, but there are relatively few actual field observations of the speed at which salt weathering takes place on natural rock outcrops. Elsewhere in this chapter we have referred to rates of weathering on playas and in tafoni. Here we first report the results of a study that was undertaken in the Karakoram Mountains.

In the Karakoram Mountains of north-west Pakistan, salt efflorescences occur widely in the dry sub-nival zones below an altitude of about 4600 m and Hewitt (1969: 101–104) thought that salt weathering was an important process that had, in the Dumordo Valley, caused slopes to be sufficiently undermined as to cause moderately large rock falls. Goudie (1984) found that hexahydrite and gypsum were the dominant efflorescence minerals in the area and attempted to assess the degree of weathering they had caused on landforms of known age that were associated with the retreat history of glaciers during the twentieth century.

For instance, in the Hasanabad Valley some of the moraine and rock fall material (mostly igneous and metamorphic) post-dating the retreat of the Hasanabad glacier as recorded between the 1920s and 1950s (Goudie et al 1984) had been rendered crumbly and disintegrated when kicked.

More impressively, in the Minapin Valley (Figure 6.18), efflorescences derived from the evaporation of melt water seeping from the glacier have led to the widespread disintegration of morainic boulders that have been deposited on a rock surface that was exposed at some time between 1954 and 1961. Analysis of rock hardness using the Schmidt hammer indicated a substantial difference in rock hardness between relatively unweathered till boulders on the present Minapin Glacier surface and those laid down in the mid- to late 1950s (Table 6.4) and subsequently softened by salt weathering. The weathered rocks have a mean value of around 35 (slates) and 33 (volcanics and schists), while the unweathered materials for the same rock categories both have values of 47.

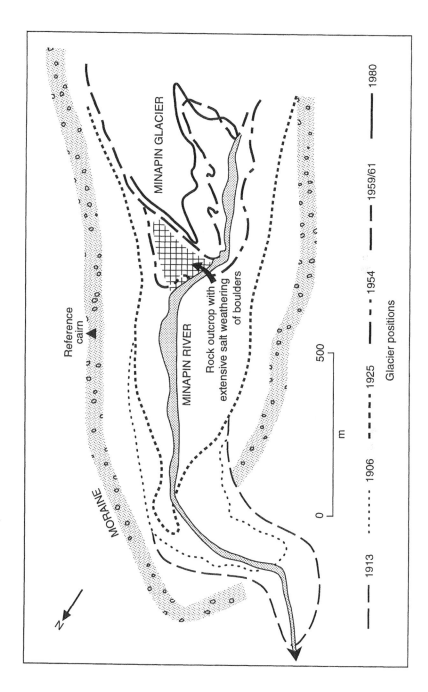

Figure 6.18. Location of severely weathered till on the margins of the Minapin Glacier, Karakoram Mountains (from Goudie, 1984: figure 1). Reproduced with permission of Cambridge University Press

Table 6.4. Schmidt hammer readings on Minapin Tills

Unweathered till boulders on present day Minapin Glacier		Weathered till boulders from post-1954 moraine	
Slates	Volcanic rocks and schists	Slates	Volcanic rocks and schists
42.6	47.1	27.2	35.6
48.0	44.7	26.9	28.5
45.0	52.2	39.2	28.5
45.5	43.0	32.8	37.4
49.5	49.4	36.5	38.5
49.9		46.8	26.8
50.4	$x=47.28$	41.8	
41.7		42.2	$x=32.55$
49.5		41.1	
50.0		38.5	
48.7		26.2	
		39.9	
$x=47.35$		37.6	
		37.7	
		32.5	
.		36.8	
		40.6	
		36.8	
		$x=36.27$	

An attempt to gauge rates of salt weathering over a longer time was made by Amit et al (1993). They investigated the development of shattered gravels and regolith soils in the Negev Desert of Israel. They found that the rate of the shattering process is not linear but exponential, and becomes asymptotic with time. As Figure 6.19 shows, the first stages of regolith development and gravel shattering are very rapid, with as much as 70% of the gravel shattering in the soil horizon being accomplished in no more than 14 000 years. The achievement of 80% shattering takes a much longer period of time — perhaps as much as 500 000 years. Amit et al (1993) account for this pattern of development in the following way. At first the presence of a very gravelly and permeable profile allows even fairly modest amounts of salts (halite and gypsum) to cause shattering. However, with time, sealing of the soil surface occurs because of the development of a pavement and the accumulation of dust and gypsum. The plugging of the soil serves to change the wetting depth, temperature and moisture regimes of the soil profile and at this stage of soil development the shattered gravel horizon is no longer subject to the extreme

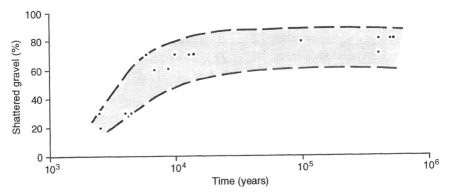

Figure 6.19. Progress of gravel shattering with time in the regolith of the Negev Desert, Israel, as a result of salt weathering. Modified from Amit et al (1993). Reproduced with permission of Elsevier Science

conditions of wetting and drying that promote haloclasty. The salts tend to cement the gravel instead of shattering them.

Matsuoka (1995), working in the arid Sør Rondane mountains of Antarctica, noted that older geomorphic surfaces affected by salt weathering had progressively finer debris. He also emplaced some soft tuff blocks that had previously been soaked with various saline solutions. After four to five years the blocks treated with pure water and gypsum solution showed scant visible signs of breakage. By contrast, blocks that had been soaked with sodium sulphate were considerably cracked and rounded, and those that had been soaked with sodium chloride were completely broken down.

Finally, Goudie et al (in press) emplaced pre-weighed rectangular blocks of Jurassic limestone (Bath Stone) on a pavement surface in the coastal Namib Desert for a period of two years. Some of the blocks suffered from extensive disintegration over that short time period, and geochemcial studies showed that the cause was the uptake of sodium chloride from the desert surface. It was argued that frequent cycles of wetting (salt solution) and drying (salt crystallisation) associated with the frequent fog events in the area were responsible for the rapid rates of attack.

There are considerable difficulties involved in measuring rates of salt weathering, especially in the (probably usual) situations where salts act in tandem with a range of other weathering processes. Several estimates have been made in a more indirect manner to the studies reviewed here, where the development of tafoni and alveoles on dated surfaces has been taken to be a measure of the rate of salt weathering. As we have seen earlier in the chapter, however, the role of salts in the development of such features has by no means been conclusively established.

CONCLUSIONS

The discussion undertaken in this chapter has shown that the geomorpho-logical role of salt weathering spans a range of scales from the cracking of individual sand grains, through the production of a range of micro-weathering forms, to the development of large overhangs, oversteepened cliffs, large desert depressions and even extensive erosion surfaces. The examples given have also shown that salt weathering contributes to landscape development in polar, coastal, desert, Mediterranean and seasonal tropical environments.

It is also clear from the material reviewed in this chapter that considerable debate still remains about exactly how salt contributes, and how important such contributions are compared with the action of other processes. Surprisingly, the formation of even fairly small and simple landforms has proved hard to explain in any general way, and there appears to be huge variations in process–form links in different environments.

7 Diagnosing, Solving and Managing Salt Weathering Hazards

INTRODUCTION

Having discussed the causes and consequences of salt weathering, we conclude this book in a more positive light by reviewing some of the methods that can be employed to diagnose, avoid and reduce the effects of the salt weathering hazard for buildings, engineering structures and historic monuments. In the same way that we began the book with some case studies of the deleterious impacts of salt weathering (Chapter 2), we end with some further case studies of managing the salt weathering hazard.

In general terms we can categorise these methods into a series of groups:

1. Diagnose the salt weathering hazard and evaluate its severity.
2. Locate new structures away from zones with the environmental ingredients that can cause salt weathering, or move vulnerable objects to safer ground.
3. Reduce the severity of environmental factors that can cause salt weathering
     by reducing groundwater levels
     by reducing air pollution
     by modifying micro-climate conditions
     by the use of salt free materials in aggregates, etc.
4. Use conservation and preventative solutions on the buildings and structures themselves
     use salt- and moisture-retarding membranes
     leach out salts
     use surfactants
     impregnate stonework
     use salt-resistant materials
5. Do nothing.

These methods should not necessarily be seen as mutually incompatible (apart from 5, perhaps, which is a valid response in many cases!), nor should they be seen as one-off solutions to the problem. Salt weathering problems cannot be solved instantaneously. Whatever is done about salt weathering, there needs to be a critical assessment of its success, and serious long-term management of the situation. Deciding what needs to be done once a salt weathering problem has been diagnosed requires an appreciation of the scientific and economic factors, as well as aesthetic and cultural considerations in many cases. The management strategy followed, for example, on a salt-affected historical monument might be different from the strategy adopted for a modern concrete bridge in identical circumstances. Expert systems for aiding the appropriate management solutions for deteriorating buildings and structures are currently being designed (e.g. the Masonry Damage Diagnostic System developed by Van Hees et al, 1996), which will help with such management decisions.

In managing salt weathering problems there are a whole range of techniques and products which can be used, and these are themselves subjected to continuous development and testing. Commonly, for example, consolidants and water repellents are applied to test specimens of stone and materials and then subjected to salt crystallisation tests (see Chapter 4) and other environmental simulations. Moropoulou et al (1992), for example, report salt crystallisation testing to assess the performance of consolidants on calcareous sandstones in the island of Rhodes. Conservators and scientists are working closely to improve the realism of such simulations, bearing in mind the problems with experimental tests discussed in Chapter 4. Baronio et al (1992) discuss one interesting approach, which involves the use of physical models (i.e. test walls) in the open air to test the impacts of consolidants on salt-affected walls.

DIAGNOSING THE PROBLEM

It is vital, before any management strategies are devised and followed, that any salt weathering problems are correctly diagnosed and understood. Several approaches can be followed, starting with a visual assessment of the problems, through *in situ* non-destructive testing, to detailed laboratory investigations. Fitzner et al (1995) have developed a scheme for the visual classification and mapping of weathering forms developed on monument and building surfaces which includes the identification of features diagnostic of the presence of salt and weathering activity (as shown in Table 7.1). Such methods can provide the basis for an objective assessment of the degree of damage to a stone or materials surface. Fitzner et al (1995) spilt deterioration features into four main types: loss of stone material, discoloration/deposits, detachment and fissures/deformation. As can be seen from Table 7.1, many of the examples of such features can be related to some form of salt attack. Although the visual assessment of damage is useful and can produce helpful illustrative "damage maps", we need to probe beneath the surface to assess the nature and causes of damage more fully.

A range of non-destructive *in situ* test methods has been developed to find out more about the non-visible progress of stone decay. Thus, for example, ultrasonic measurements can provide useful information about the moisture content and porosity of subsurface stone, although considerable care needs to be taken in interpreting the results. In addition, measurements can be taken of

Table 7.1. Classification of weathering forms, focusing on those related to salt problems. Adapted from Fitzner et al. (1995)

Feature	Description
Loss of stone material	
Back-weathering (uniform back-weathering parallel to original stone surface), e.g.	
Back-weathering due to loss of crusts	*Uniform back-weathering parallel to original stone surface, related to loss of crusts and adhering stone material*
Relief (morphological change of the stone surface related to partial or selective weathering), e.g.	
Alveolar weathering	*Production of honeycombed surface*
Break-out (loss of compact stone fragments)	
Discoloration/deposits	
Discoloration (alteration of the original stone colour)	
Soiling (dirt deposits on the stone surface), e.g.	
Soiling by pollutants from the atmosphere	*Poorly adhesive, mainly grey to black deposits of dust, soot, fly ash, etc. on stone surface*
Loose salt deposits (poorly adhesive deposits of salt aggregates), e.g.	
Efflorescences	*Poorly adhesive deposits of salt aggregates on the stone surface*
Subflorescences	*Poorly adhesive deposits of salt aggregates below the stone surface. Frequently in the zone of detachment of scales*
Crust (firmly adhesive deposits on the stone surface) e.g.	
Dark coloured crust tracing or changing the stone surface	*Compact deposits, grey to black coloured Mainly due to deposition of pollutants from the atmosphere, gypsum common*
Light coloured crusts tracing or changing the stone surface	*Compact deposits, light coloured, mainly due to precipitation processes—salt, calcite or silicate materials common*
Coloured crusts tracing or changing the stone surface	*Compact coloured deposit, mainly due to precipitation processes, salt or iron/ manganese common*

Continued

Table 7.1. *continued*

Feature	Description
Biological colonisation (by micro-organisms and higher plants)	
Detachment	
Granular detachment (detachment of grainy stone particles, individual grains or grain aggregates)	*May sometimes be associated with salt presence in stone*
Crumbling (detachment of larger, compact stone elements in the form of crumbs)	*May sometimes be associated with salt presence in stone*
Splintering	*May sometimes be associated with salt presence in stone*
Flaking	*May sometimes be associated with salt presence in stone*
Contour scaling	
Detachment of stone elements depending on stone structure	
Detachment of crusts with attached stone material	*May sometimes be associated with salt presence in crusts*
Fissures/deformation	
Fissures dependent and independent of stone structure	
Deformation (especially of marble slabs)	

the microenvironmental conditions faced by building walls and other surfaces to help the interpretation of the damage problems. Thus, for example, Baggio et al (1993) combine field measurements of temperature and humidities on the outside and inside of salt-affected walls on a church in Venice with laboratory observations of the hydrological behaviour of the building materials used, to feed into a numerical model to investigate the drying behaviour of the walls. This study also indicates the need for laboratory investigations to determine the geochemistry of salts present, as well as the geochemistry and porosity of the building materials themselves.

A final factor which is of great importance to diagnose is the history of treatment/management of a building or structure which may have influenced the presence or behaviour of salts. Such information can sometimes be derived from archival evidence where good records exist of past management, or can be deduced from detailed visual or laboratory observations. A good example is provided by the investigations of Theoulakis and Beloyannis (1991), who found that restorations of the Citadel of Lindos on the Island of Rhodes in the 1930s used Portland cement rich in soluble sulphates, which have caused extensive damage to the surrounding stone.

AVOIDANCE AND ZONING

One of the most effective ways to cope with salt attack is not to build structures in aggressive areas. This is particularly true of those situations where the groundwater level and salinity are the crucial controls, as in the Middle East's low-lying coastal cities. Jones (1980) summarises some of the work undertaken by British geomorphologists to produce hazard zone maps of value to both engineers and planners. In the Middle Eastern studies the potential intensity of the salt weathering hazard is basically a function of the elevational relationship between the ground surface and the limit of capillary rise, and the salinity of the rising water. As Jones points out (1980: 5), and as is shown diagrammatically in Figure 7.1, a whole range of situations can be envisaged where saline water-tables lie at or close to the ground surface, and the potential for capillary rise is well above this level, to where the water-table is so deep within the ground that the limit of capillary rise lies below even the deepest foundations. The figure shows four hazard zones and shows the capillary fringe limit as a particularly important boundary. This limit can often be identified on aerial photographs because of the presence of a well-defined tonal boundary. In addition, the zones have to be identified by gathering and analysing data on ground surface elevation, the depth to the water-table, the nature and distribution of different types of surface material (to facilitate the determination of the potential height of capillary rise), and spatial variations in the salinity of the groundwater. The four zones will be subject to change over time so that some form of long-term monitoring of variations in water-table configuration and chemistry in response to climate

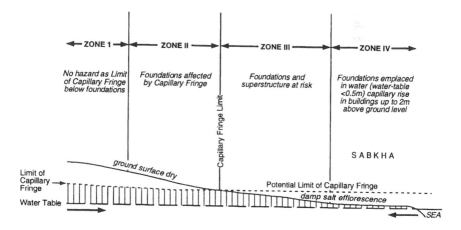

Figure 7.1. Zonal divisions and terminology (see also p. 84) used in geomorphological investigations of aggressive ground in the Middle East (modified from Jones, 1980: figure 5). Reproduced with permission

fluctuations, urban development and tidal conditions is desirable. Such long-term monitoring is seldom practised or available (Cooke et al, 1982: 175).

REDUCTION IN GROUNDWATER LEVEL

Given that the groundwater level is such an important control of salt attack (e.g. in the coastal cities of the Middle East or at Mohenjo-Daro in Pakistan), efforts need to be made either to keep the groundwater levels from rising (e.g. by the control of irrigation developments) or, if they have already reached critical levels, to make them fall. The latter of these two strategies is probably best approached by pumping of groundwater via tube-wells and its evacuation in drainage canals (disposal channels). This was the approach adopted in the Master Plan for Mohenjo-Daro (van Lohuizen-de Loeuw, 1973), where rings of tube-wells were proposed (Figure 7.2). The cost of the scheme is enormous and given the large quantities of sodium sulphate and other salts that have already permeated the ruins it is by no means certain that this will stop the problem by itself.

A total of 26 pumps has now been installed on a canal dug round the perimeter of Mohenjo-Daro, with the object of eventually lowering the water-table to 10 m below the surface. It is reported that by the end of the first year after these pumps had been installed, the level had already been lowered to approximately 6 m (Jansen, 1996).

MODIFYING LOCAL ENVIRONMENTAL CONDITIONS

Although blanket reductions of air pollution and other radical changes to urban environmental conditions are unlikely to be carried out simply to reduce salt damage to a building, such damage can provide additional support for plans to improve urban environmental conditions. In cities or areas where highly valuable components of our cultural heritage are suffering from salt hazard, costing the damage can help persuade governments to reduce pollution emissions (where possible) and improve other facets of the environment. In Oxford, for example, the City and County Councils added in estimates of costs for clearing, repainting and repairing building façades polluted by heavy traffic into their analysis of the costs and benefits of reducing the inner city traffic flows (Jarman, 1994).

In most instances, unless there is a large assemblage of historic monuments in a small area, a more realistic approach is to modify microenvironmental conditions in the immediate vicinity of a threatened building or structure. Constructing a protective "house" around an object is no automatic solution, however, and the new microenvironmental conditions need careful monitoring.

STONE TREATMENT

The third category of approach to dealing with the salt weathering hazard is to employ various categories of material treatment technique. These are

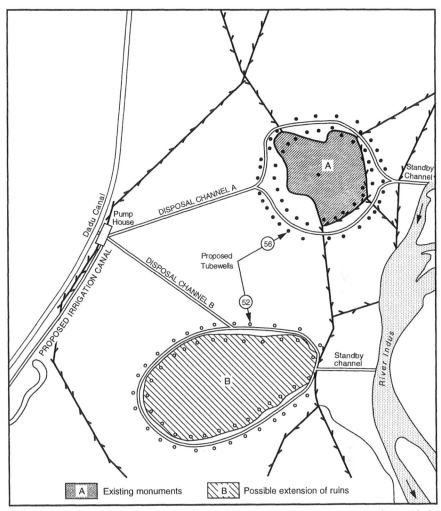

Figure 7.2. Site map of Mohenjo-Daro, Pakistan, showing areas A and B and the planned locations of tube-wells and drainage canals designed to reduce salt attack. Modified from Goudie (1977: figure 5) and reproduced with permission

discussed in detail by Amoroso and Fassina (1983), and they classify remedial action in terms of the following sequence (pp. 224–225):

Diagnosis consists of an in-depth study of the causes and mechanism of the decay processes and the history of the object in need of restoration.

Cleaning is the mechanical, chemical and physical removal of weathering crusts and dust deposited on the surface of stones.

Preconsolidation is a superficial consolidation of the stone and is applied before cleaning in cases of advanced decay where cleaning would cause considerable irreversible loss of stone.

Consolidation is the in-depth treatment of stone that has lost its cohesion to such a degree that its physical survival is imperilled. It consists of the impregnation of the weathered stone, as well as a substantial part of the underlying sound layer of the stone.

Surface protection consists of the application on unweathered stone of a superficial film which acts as a barrier towards atmospheric pollutants and rainwater. The surface layer is generally applied after consolidation to extend its effectiveness.

Reconstitution is the assembly of parts of old, consolidated stone by means of adhesives, or even of substitute parts made of new artificial stone.

Frequently, the term preservation is used alternatively to conservation to indicate any action to prevent the access of water and to render the material impregnable to pollutants.

Maintenance is the periodic inspection of stone monuments to assess the state of conservation and in particular to check the efficiency of protective treatment.

The use of chemical treatments to enable consolidation and surface protection has a long history, but is frequently not successful. Arkell (1947: 165) wrote of two particular processes and noted their lack of effectiveness

Various chemical treatments have had a vogue but all have proved fruitless in the long run. The most popular were the fluate and baryta processes. The fluate process consists of spraying the stone with fluorosilicate solutions with the object of forming a protective skin of chemically inert material like glass. The baryta process consists of spraying the decayed stone with a solution of baryta (barium hydroxide) in order to convert the soluble calcium sulphate. Other processes seek to deposit silica or insoluble soap in the pores of the stone, or to waterproof it with waxes, resins, or oils. The weakness of all these treatments is that they do not penetrate deep enough and decay goes on behind the protected skin, the ultimate effect then being worse than if no treatment had been attempted. At the best the preservative wears off in a few years and has no effect whatever.

Arkell was more favourably disposed towards lime washes, which could act as an insulating or buffering agent against harmful gases. Today there are many chemical products and techniques which can help stone conservators deal with salt-damaged materials, including consolidants (to strengthen weak materials), water repellents (to prevent water penetration and thus further salt damage), and cleaning agents (to remove crusts, efflorescences and subflorescences). We discuss some of the major issues involved in the use of such products and techniques on salt-affected structures in the following paragraphs. More detailed discussion of a wide range of stone treatments is given in Amoroso and Fassina (1983).

There are now a large number of treatments available for the consolidation of weathered materials and for water repellency, including a whole range of polymers (see, for example, Littmann et al, 1993). Their performance in the presence of high salt contents can be a problem, and we have also to consider their adhesion qualities, longevity, reversibility, depth of penetration, discoloration effects, and the use of toxic solvents. Indeed, the criteria for defining the ideal stone preservative are many (Amoroso and Fassina, 1983: 244)

It must:

- penetrate easily and deeply into the stone and remain there on drying
- not concentrate on the surface so as to form a hard crust, but must, at the same time, harden the surface sufficiently to resist erosion
- prevent moisture penetration and, at the same time, allow moisture to escape
- not alter the natural appearance of the stone (e.g. by discoloration)
- expand and contract uniformly with the stone so as not to cause flaking
- be non-corrosive and harmless in use
- be economical
- be able to maintain its preservative effect indefinitely
- be applicable under a range of moisture and temperature conditions
- be resistant to attack by acid or alkaline substances
- not take on a crystalline state and thus itself introduce the danger of surface decay owing to crystal growth.

In general, consolidants should not be applied to zones where there are major moisture movements, or to surfaces which have heavy concentrations of soluble salts unless they have first received some cleaning or poulticing.

Water repelling or hydrophobic substances, by excluding moisture, can be a successful means of reducing stone decay, but the application of water-repelling substances can sometimes exacerbate decay. As Ashurst and Ashurst (1988: 92–94) point out:

> This can happen as a result of water containing salts in solution evaporating from behind the treated layer, leaving salt crystals in the pores behind the treatment. Repeated crystallisation cycles can then lead to disruption and spalling of the treated surface.

De Witte and Bos (1992) also claim that water repellents should not be used on ferruginous sandstones which are contaminated with soluble salts, as their use may cause the impregnated layer to split off. In their view, however, consolidants do not cause problems on such salt-affected stones. Nonetheless, substances such as silica sols, which are water-based, non-toxic and economic, appear to give satisfactory results when applied to porous rocks with soluble salt contamination (Kozlowski et al, 1992a).

Various workers have evaluated the performance of water-repellent treatments and consolidants in the presence of salts by using laboratory simulation tests. Villegas and Vale (1992), for example, established that organosilica products (e.g. BS28 produced by Wacker) were relatively durable in accelerated weathering tests, whereas acrylic products (e.g. Paraloid B72 produced by Röhm and Haas) were not.

Ciabach (1996) tested the effects of soluble salts on sandstones impregnated with silicone microemulsions, which act as water repellents and strengtheners. He found that such microemulsions (e.g. Wacker 550) were unstable in the presence of water-soluble salts, as two distinct fractions are formed (one resin-rich and one watery).

Cleaning of masonry and the removal of soluble salts is a fundamental way of reducing the rates of stone decay. Lasers are now being used increasingly as a means of cleaning building materials (Orial and Riboulet, 1993) and for getting rid of black gypsiferous encrustations (Maravelaki et al, 1992). However, a more common means of attempting to cope with the effects of salt weathering is the removal of soluble salts with, for example, nebulised water treatments. These can sometimes be combined with chemical treatments that involve the application of solutions that can lead to the increased mobility of deleterious salts. An example of this is the use of ammonium carbonate to mobilise sparingly soluble gypsum. This treatment is based on the following formula:

$$CaSO_4 \cdot 2H_2O + (NH_4)_2CO_3 \rightarrow (NH_4)_2SO_4 + CaCO_3 + 2H_2O$$

The method is assessed by Alessandrini et al (1993).

Similarly, salts can sometimes be removed by the application of poultices, including those composed of bentonite clays (Barbosa et al, 1993), attapulgite, or ethylenediaminetetraacetic acid (EDTA) (DeWitte and Dupas, 1992). When clay poultices are used for "desalination" the wall is first sprayed with mists of clean water for a period of a few days. When this wetting process is complete, the absorbent clay (or diatomaceous earth) is added to enough clean water to make a soft, sticky paste. The poultice is plastered onto the wetted wall and may be kept in place with wire or hessian. As the poultice dries out it draws salt-laden water from the masonry, though the process may need to be repeated more than once if salt levels are going to be reduced to an acceptable degree (Ashurst and Ashurst, 1988: 69–71). Other workers have used biological poultices containing urea, glycol and water to remove black, gypsum-rich crusts on Pentelic marble and compared it with the use of EDTA (Kouzeli, 1992). EDTA is not a highly desirable substance to use to remove gypsum from calcareous stone as it indiscriminately removes gypsum and calcite. The biological poultice proved less damaging, but equally effective.

An alternative technique for the treatment of salt-contaminated masonry is to use a sacrificial render. The principal behind this is that a porous render is applied to the wall and the evaporation of moisture from the wall results in soluble salts being transferred from the masonry to the render. The render may deteriorate through time, but it serves to protect the masonry itself against continued decay, particularly where rising damp is a problem (Ashurst and Ashurst, 1988: 71–72).

HUMIDITY MODIFICATION

Humidity changes can cause salt weathering to occur, so that the control of relative humidity can be an important means of controlling the decay of materials. As Price and Brimblecombe (1994: 90) have pointed out

Crystallisation and hydration occur only at particular relative humidities, and damage can therefore be prevented by keeping the ambient relative humidity above or below these critical values.

They argue (p. 91)

Preventive conservation is straightforward in principle: the ambient RH needs to be controlled so that it never coincides with the equilibrium RH. If the ambient RH is permanently above the equilibrium RH, the salt will remain in solution at all times. If on the other hand the ambient RH is permanently below the equilibrium RH, the salt will remain solid at all times.

The critical relative humidities for single salts at particular temperatures are well known (Table 7.2). So, for example, material containing sodium chloride should not be allowed to move back and forward across a RH value of 75% at 20°C. Humidity should preferably be kept below that critical value so that the salt remains in the solid state.

However, it is rare for a monument to be contaminated with a single salt, and if salt crystallisation involves a range of saturated solutions, the situation becomes much more complex than for single salts. A "safe" range of relative humidities cannot simply be predicted on the basis of the relative humidities at which crystallisation and hydration occur in the component salts in isolation. Fortunately, Price and Brimblecombe (1994) have used Pitzer's virial methods to calculate the relative humidity of air in equilibrium with any given mixed salt solution. This means that a safe range of relative humidities can be predicted providing that a full analysis is available of all the ions present. Another useful study of critical temperature and humidity conditions for simple mixtures of salts is provided by Steiger and Zeunert (1996).

PREVENTION OF DAMAGE TO BITUMEN PAVEMENTS

As was noted in Chapter 2, salt can be a major cause of damage to bitumen roads and runways (see, for example, Januzke and Booth, 1984). The main means of attempting to reduce this problem have been summarised by Obika et al (1989).

Table 7.2. Equilibrium relative humidities at 20°C

Sodium sulphate	94%	Sodium nitrate	75%
Potassium sulphate	98%	Potassium nitrate	94%
Calcium sulphate	99.96%	Calcium nitrate	56%
Magnesium sulphate	90%	Magnesium nitrate	53%
Ammonium sulphate	81%		
Sodium chloride	75%		
Potassium chloride	85%		
Calcium chloride	33%		
Magnesium chloride	34%		
Ammonium chloride	80%		

The first aim should be to keep the salt content of compacted basecourse and subgrade to low limits, and Table 7.3 shows some suggested limits developed by Fookes and French (1977) and reproduced by Obika et al (1989: table 3).

However, it may not always be possible or economically feasible to avoid the use of saline materials and so other procedures may be necessary. One partial solution is to use a thick and dense asphalt, with the intention of greatly reducing the evaporative loss of moisture and hence the migration and crystallisation of salt at the surface. Additionally, certain types of bituminous prime may help to reduce permeability, while salt migration by the upward capillary rise of soil moisture can be prevented by placing an impermeable or semi-permeable membrane in the base course. It is also important to use aggregates of high durability and soundness.

PROTECTION OF BRIDGES AND CONCRETE STRUCTURES AGAINST SALT ATTACK

Mallett (1995, chapter 5) presents a useful review of some of the methods that are available to deal with salt attack on bridges, while a more general review of the means of dealing with the effects of chlorides and sulphates on concrete structures more generally is provided by Bijen (1989). Important measures include waterproofing and good drainage and care not to over-apply de-icing salts on bridges. Various concrete sealing materials are available to control chloride intrusion, including epoxies, methacrylate, urethane and silane. It is also possible to remove salt from concrete by electrochemical means, while cathodic protection is a well-established anti-corrosion method for protecting steel reinforcements that may be exposed to aggressively saline environments. A full discussion is, however, beyond the scope of this book.

CASE STUDIES

St Maria dei Miracoli Church, Venice

Like many buildings in the historic centre of Venice, the Church of St Maria de Miracoli has been badly affected by salt efflorescences on the walls. The church, built between 1481 and 1488, is decorated inside and out by marble slabs (of various types) which completely cover the underlying brick masonry. This beautiful church has been the subject of long-term restoration work. In the late 19th century several marble slabs were replaced with new ones which were affixed with Portland cement. Continued problems with moisture migrating up the walls lead to the insertion of a damp course in the 1960s.

Despite these remedial measures, salt efflorescences are still a major problem. Recent studies by Fassina et al (1996) indicated that the Portland cement has encouraged the migration of salts from bricks to marble slabs, and although the damp course has prevented further salt water coming in from

Table 7.3. Suggested limits of salt in aggregates (salts expressed as weight percentage anion of dry aggregate) by Fookes and French (1977)

	Bituminous wearing course/basecourse		Unbound base under		Unbound sub-base under	
	Thick/dense (A)	Thin/porous (B)	(A)	(B)	(A)	(B)
Soluble chloride anions in the aggregate, in possible presence of groundwater with very low dissolved salts	0.5 (0.2)	0.4 (0.1)	1.0 (0.5)	0.4 (0.1)	1.0 (0.5)	0.4 (0.1)
Soluble anions in the aggregate where more than 10% anion is sulphate, in possible presence of groundwater with very low dissolved salts	0.3 (0.2)	0.2 (0.1)	0.3 (0.2)	0.3 (0.1)	0.5 (0.4)	0.2 (0.1)
If local groundwater contains salts in solution	The design must also minimise moisture migration in road section	The design must not allow capillary moisture to reach the road surface				
Soluble chloride anions in the aggregate in possible presence of transient moisture (rain, dew, etc.) only	0.6 (0.3)	0.5 (0.2)	2.0 (0.8)	1.0 (0.4)	3.0 (0.2)	
Soluble anions in the aggregate where more than 10% anion is sulphate, in possible presence of transient water (rain, dew, etc.) only	0.5 (0.2)	0.3 (0.1)	0.5 (0.2)	0.3 (0.1)	2.0 (1.0)	

Figures in parentheses refer to soft or friable aggregates of high hazard potential, e.g. clayey or chalky limestones (say water absorption (BS812)>5%; soundness loss (ASTM C88) by $MgSO_4$>30: ACV (BS812)>25).

Groundwater with very low dissolved salts is arbitrarily defined as containing less than 0.1 total dissolved solids.

The thick/dense surface is assumed to be a non-cement bound asphalt at least 40 mm thick.

Moisture movement in the road section from groundwater sources can be minimised by raising the road on an embankment, the use of an impervious surface course, the use of an impervious membrane (e.g. heavy duty polythene) or sand/bitumen layer or prime coat in the lower part of the road.

below, it has not stopped the action of soluble salts already impregnating the wall.

Detailed microenvironmental studies by Baggio et al (1993) helped develop a numerical model of the situation to prevent further damage. The model indicates that leaving a cavity between the marble slabs and the brickwork should protect the marble from further salt damage. As a precursor, it was decided to remove and desalinate the marble slabs where necessary. Laboratory analyses of the salt efflorescences showed high levels of sulphate (as aphthitalite and thenardite), as well as gaylussite and trona, which may have originated from the Portland cement.

Salts were removed from the marble slabs by soaking in deionised water, after experimental trials to establish the best protocol. The combination of understanding the management history of the building, diagnosing the major types and sources of salts and moisture and scientific evaluation of the best solutions has led to a good understanding of the problems and the best way to tackle them.

Twelfth century frieze, Lincoln Cathedral, UK

A difficult management challenge is posed by the 21 panels of a twelfth century frieze embedded into the ashlar of the West Front of Lincoln Cathedral (Figure 7.3). The frieze depicts various scenes from the Bible and, according to Larson (1992: 1168), represents: "the most important sculpture ensemble of the type in England". Lincoln Cathedral sits on a hill top and has been badly affected by atmospheric pollution from a range of local and distant sources. The frieze has become badly disfigured by a sooty gypsum crust and by the late 1980s the cathedral authorities had been forced to cover the frieze with protective wooden covers while deciding the best conservation strategy.

Larson (1992) describes the various stages involved in conserving the frieze. Archival evidence showed that various restorations had been carried out in the past and detailed laboratory investigations helped to diagnose the state of the frieze (made of Ancaster Stone, a local ooidal limestone). The frieze was found to be made of face-bedded stone, which has aided decay as the soluble salts are able to spall off whole layers of stone. The frieze was also found to show evidence of pigments, possibly dating from the medieval period.

After the diagnosis it was decided to clean off the polluted crust, but great care was needed because of the blistered state of some areas. Air abrasion techniques were tested initially, but then laser cleaning was used as a more controllable, precise method of cleaning complex geometries. Because the stone has been found to be face-bedded, deep consolidation is not thought to be appropriate, and such consolidation would also "bury" traces of the pigments, which may be of great historical interest.

A larger dilemma is whether the entire frieze should be removed from the West Front and stored in a protective atmosphere, whilst replicas are fixed on the West Front.

Figure 7.3. The Lazarus frieze from the West Front of Lincoln Cathedral has become badly damaged by sulphation (photograph by H. A. Viles)

Tomb 825: planning restorations at Petra, Jordan

Jordanian and German scientists and conservators have set up a joint project to conserve and restore the rock-hewn Nabatean buildings at Petra documented in Chapter 2 (Figure 7.4). Tomb 825 (the Tomb of Fourteen Graves) is the first monument to be tackled through detailed documentation; architectural description, assessment of damage and the formulation of a restoration plan, as described by Aslan and Shaer (1996).

Scaffolding of the tomb permitted detailed investigation. Basic documentation of the monument was carried out using a total station (electronic distance measurement and theodolite and computer), and then detailed observations of the architectural features were made. Mapping of the weathering features was carried out following the methodology of Fitzner et al (1995). Back-weathering was a major type of weathering phenomenon on Tomb 825,

Figure 7.4. Restoration of monuments at Petra, Jordan involves Jordanian and German scientists and requires careful mapping of weathering features before any treatments are undertaken (photograph by H. A. Viles)

coupled with alveolar weathering. Rising damp, salts and the inherent vulnerability of limonite bands within the sandstone are all inferred to be causing decay here.

Following this assessment of the condition of Tomb 825, restoration plans have been developed. Aslan and Shaer (1996) recommend that major work is only required on the upper half of the façade. In this portion, new material should be used to rebuild the protective cornice, new mortars injected to fix the exfoliated layers, consolidants used to repair flaking areas and salt crusts and deposits removed. Current work is being carried out to select the most appropriate materials.

Removing gypsum from terracotta decorations, Schloss Schwerin, northern Germany

Ornate terracotta decorations on Schloss Schwerin added in the mid-19th century have become blackened and encrusted with gypsum as a result of air pollution. Wittenburg et al (1996) report investigations of the state and causes of deterioration and attempts to discover a good solution to the problem. Firstly, the deterioration phenomena were mapped and investigations of material strength carried out. Measurements of salt contents showed 3–10% by weight of gypsum concentrated in the first few millimetres. Studies were also made of the capillary uptake capacity of the terracotta to see whether stone consolidants would be taken up adequately. Then two methods to remove gypsum were tested. The first involved a water poultice and the second used a poultice of ammonium carbonate, which converts gypsum to calcite and easily soluble ammonium sulphate through the reaction

$$CaSO_4 \cdot 2H_2O + (NH_4)_2CO_3 \rightarrow CaCO_3 + (NH_4)_2SO_4 + 2H_2O$$

Detailed testing of these methods on ten terracotta decorations showed that the effectiveness of water poultices was highly limited by the low solubility of gypsum. The ammonium carbonate poultices were particularly effective, and although care needs to be taken to remove all reaction products, this method seems to have potential in this case.

Protecting salt sculptures, Wieliczka salt mine, Poland

An unusual example of the salt weathering hazard involves the impact of air pollution on underground sculptures made of rock salt in the Wieliczka salt mine, recognised as a Unesco world heritage site. The mine, near Krakow in southern Poland, has been worked on as a source of rock salt since the 13th century. There are now over 200 km of underground passages connecting a whole series of excavation chambers. Miners have, over the centuries, created entire underground chapels and carved huge numbers of statues. Up to one million visitors a year come to the mine, and one of the chambers houses a sanatorium for people with respiratory problems. In recent years many of the statues have started to dissolve away, and several are becoming discoloured. A joint American and Polish team of scientists has started to monitor the environmental conditions in the mine, assess the problems and suggest solutions (Salmon et al, 1996).

Measurements have been taken of gaseous and particulate air pollutants. Sulphur isotope analyses have been used to fingerprint sources of sulphur and thermodynamic calculations undertaken to see whether the pollutants deposited on the surface of sculptures are capable of reducing the relative humidity of deliquescence, and thus increasing the weathering of the salt sculptures.

Levels of sulphur dioxide outside the mine are up to 40 ppb in winter, and only up to 0.5 ppb inside as most of the sulphur dioxide is lost to interior surfaces. Similarly, the annual outdoor coarse particle concentrations are about $30\,\mu g\,m^{-3}$ and outdoor fine particle concentrations are about $50\,\mu g\,m^{-3}$, whereas indoor concentrations are about 20 and $29\,\mu g\,m^{-3}$, respectively.

Sulphur isotope analyses confirmed that sulphur from outdoor air pollution was being taken up by surfaces inside the mine. The nitrate concentrations on surfaces within the mine are luckily fairly low, as the low deliquescence point of nitrates would exacerbate the salt weathering problems. Thermodynamic calculations and experiments on polluted sculptures show that pollutant deposition has reduced the effective relative humidity at which deliquescence, and thus removal of a liquid layer, can occur by 1% or more. Attempts to protect and conserve these sculptures must take into account these findings.

CONCLUSIONS

Ultimately, the main necessity for dealing with the attack of salt on human-made structures is awareness — awareness of such issues as where salt is a problem, which salts are particularly aggressive, which climatic conditions are particularly conducive to salt weathering, which materials are especially vulnerable, and which mechanisms are involved in different situations.

The case studies presented in this chapter, all of which deal with buildings, monuments or objects of cultural and historical significance, illustrate some possible approaches to diagnosing, solving and managing the salt hazard. It is clear that there are no easy solutions to such a complex and insidious series of problems, but there is now available a huge variety of techniques, materials and scientific ideas with which to approach them. As environmental conditions continue to change over a range of time-scales it is essential that responses to the salt weathering hazard are not seen as "quick fix" solutions, but rather as long-term environmental management challenges.

LIBRARY, UNIVERSITY OF CHESTER

References

Abd el Hady, M. M. (1992) Weathering and disintegration of monumental marble in Egypt. In: *Proceedings, 7th International Congress on Deterioration and Conservation of Stone* (Eds J. D. Rodrigues, F. Henriques and F. Telmo Jeremias). Lisbon: Laboratorio Nacional de Enghenaria Civil, 139–150.

Abele, G. (1983) Flachen hafte Hanggestaltung und Hangzerschneidung im Chilenisch — Peruanischen Trockengebiet. *Zeitschrift für Geomorphologie Supplementband* **48**, 197–201.

Abrahams, A. D. and Parsons, A. J. (1994) *Geomorphology of Desert Environments.* London: Chapman and Hall.

Addleson, L. and Rice, C. (1991) *Performance of Materials in Buildings.* Oxford: Butterworth-Heinemann.

Aires-Barros, L. and Mauricio, A. (1996) Chronology, probability estimations and salt efflorescence occurrence forecasts on monumental building stones surfaces. In: *Proceedings, 8th International Congress on Deterioration and Conservation of Stone* (Ed. J. Riederer) Berlin: Ernst und Sohn, 497–511.

Akiner, S., Cooke, R. U. and French, R. A. (1992) Salt damage to Islamic monuments in Uzbekistan. *Geographical Journal*, **158**, 257–272.

Albouy, M., Deletie, P., Haguenauer, B., Nion, S., Schreiner, E., Rewerski, J. and Seigne, J. (1993) Petra, la cité rose des sables. La pathologie des grès et leur traitement dans la perspective d'une préservation et d'une restauration des monuments. In: *Conservation of Stone and Other Materials* (Ed. M. J. Thiel). London: Spon, 376–385.

Albritton, C. C., Brooks, J. E., Issawi, B. and Swedan, S. (1990) Origin of the Qattara Depression, Egypt. *Bulletin of the Geological Society of America* **102**, 952–960.

Alcade, M. and Martin, A. (1988) Macroscopical study of the stone alteration I. The Cathedral of Seville. In: *Proceedings, 6th International Congress on the Deterioration and Conservation of Stone.* Torun: Nicolas Copernicus University Press Department, 216–225.

Alessandrini, G., Tonioli, L., Antonioli, A., Di Silvestro, A., Piacenti, R., Righini Ponticelli, S. and Fornice, L. (1993) On the cleaning of deteriorated stone minerals. In: *Conservation of Stone and Other Materials* (Ed. M.-J. Thiel). London: Spon, 503–511.

Al-Izz, M. S. A. (1971) *Landforms of Egypt.* Cairo: American University of Cairo Press.

Allchin, B. and Allchin, R. (1968) *The Birth of Indian Civilisation.* Harmondsworth: Penguin, 365.

Amit, R., Gerson, R. and Yaalon, D. H. (1993) Stages and rate of the gravel shattering process by salts in desert reg soils. *Geoderma* **5**, 295–324.

Amoroso, G. G. and Fassina, V. (1983) *Stone Decay and Conservation.* Materials Science Monographs 11. Amsterdam: Elsevier.

André M-F. (1995) Postglacial microweathering of granite roches moutonnées in northern Scandinavia (Riksgränsen area, 68°N). In: *Steepland Geomorphology* (Ed. O. Slaymaker). Chichester: Wiley, 103–127.

Arad, A. and Morton, W. (1969) Mineral springs and saline lakes of the western Rift Valley, Uganda. *Geochimica et Cosmochimica Acta* **33**, 1169–1181.

Arkell, W. J. (1947) *Oxford Stone*. London: Faber and Faber.

Arnold, A. (1981) Nature and reactions of saline minerals in walls. In: *Preprints of the Contributions to the International Symposium on the Conservation of Stone, II, Bologna*, 13–23.

Arnold, A. (1984) Determination of mineral salts from monuments. *Studies in Conservation* **29**, 129–138.

Arnold, A. and Zehnder, K. (1988) Decay of stony materials by salts in humid atmosphere. In: *Proceedings, 6th International Congress on the Deterioration and Conservation of Stone*. Torun: Nicolas Copernicus University Press Department, 138–148.

Arnold, A. and Zehnder, K. (1990) Salt weathering on monuments. In: *The Conservation of Monuments in the Mediterranean Basin* (Ed. F. Zezza). Brescia: Grafo, 31–58.

Ashbel, D. (1949) Frequency and distribution of dew in Palestine. *Geographical Review* **39**, 291–297.

Ashurst, J. and Ashurst, N. (1988) *Practical Building Conservation*. English Heritage Technical Handbook, Vol. 1, Stone Masonry. Aldershot: Gower Technical Press, 100 pp.

Aslan, Z. and Shaer, M. (1996) Tomb 825: a case study in planning for restoration of a selected monument in Petra. In: *Proceedings, 8th International Congress on Deterioration and Conservation of Stone* (Ed. J. Riederer). Berlin: Ernst und Sohn, 1109–1115.

Auden, J. B., Gupta, B. C., Roy, P. C. and Husain, M. (1942) Report on sodium salts in *Reh* soils in the United Provinces, with notes on occurrences in other parts of India. *Records, Geological Survey of India* **77**, 1–45.

Ausset, P., Crovisier, J. L., del Monte, M., Furlan, V., Girardet, F., Hammecker, C., Jeannette, D. and Lefevre, R. A. (1996) Experimental study of limestone and sandstone sulphation in polluted realistic conditions: the Lausanne Atmospheric Simulation Chamber (LASC). *Atmospheric Environment* **30**, 3197–3207.

Australian Standing Committee on Soil Conservation (ASCSC) (1982) *Natural and Human-induced Salinisation in Australia*. Canberra: ASCSC [cited in M. A. J. Williams and R. C. Balling (Eds) (1996) *Interactions of Desertification and Climate*. London: Arnold].

Baggio, P., Bonacina, C., Stevan, A. G. and Fassina, V. (1993) Analysis of moisture migration in the walls of the Santa Maria dei Miracoli church in Venice. In: *Conservation of Stone and Other Materials*. Volume 1: Causes of Disorders and Diagnosis (Ed. M.-J. Thiel). London: Spon, 170–177.

Ball, D. G. and Nichols, R. L. (1960) Saline lakes and drill hole brines, McMurdo Sound, Antarctica. *Bulletin Geological Society of America* **71**, 1703–1708.

Ball, J. (1927) Problems of the Libyan Desert. *Geographical Journal* **70**, 21–38, 105–128, 209–224.

Barbosa, A. C. de F., Santiago, C. C. and De Oliviera, M. M. (1993) The use of Brazilian bentonites for cleaning purposes. In: *Conservation of Stone and Other Materials* (Ed. M.-J. Thiel). London: Spon, 550–557.

Bari, M. A. and Schofield, N. J. (1992) Lowering of a shallow, saline water table by extensive eucalyptus reforestation. *Journal of Hydrology* **133**, 273–291.

Baronio, G., Binda, L., Cantoni, F. and Rocca, P. (1992) Durability of preservative treatments of masonry surfaces: experimental study on outdoor physical models. In: *Proceedings, 7th International Congress on Deterioration and Conservation of Stone* (Eds J. Delgado Rodriguez, F. Henriques and F. Telmo Jeremias). Lisbon: Laboratorio Nacional de Enghenaria Civil, 1083–1092.

Beaudet, G. and Michel, P. (1978) *Recherches Géomorphologiques en Namibie Centrale*. Strasbourg: Association Géographique d'Alsace.

Beaumont, B. (1968) Salt weathering on the margin of the Great Kavir, Iran. *Geological Society of America Bulletin* 79, 1683–1684.

Begonha, A., Sequeira Braga, M. A. and Gomes Da Silva, F. (1996) Rain water as a source of the soluble salts responsible for stone decay in the granitic monuments of Oporto and Braga-Portugal. In: *Proceedings, 8th International Congress on Deterioration and Conservation of Stone* (Ed. J. Riederer). Berlin: Ernst und Sohn, 481–487.

Bell, F. G. (1992) *Engineering in Rock Masses*. Oxford: Butterworth-Heinemann.

Beloyannis, N. and Dascalakis, S. (1990) Marine deterioration of marbles from the Cyclades, the case of marble statues in Delos and Naxos. In: *The Conservation of Monuments in the Mediterranean Basin* (Ed. F. Zezza). Brescia: Grafo, 189–193.

Bijen, J. M. (1989) Maintenance and repair of concrete structures. *Heron (Delft)* 34(2), 1–82.

Birot, P. (1954) Désagrégation des roches cristallines sous l'action des sels. *Comptes Rendus de l'Academie des Sciences Paris* 28, 1145–1146.

Blackburn, G. and Hutton, J. T. (1980) Soil conditions and the occurrence of salt damp in buildings of metropolitan Adelaide. *Australian Geographer* 14, 360–365.

Blackwelder, E. (1929) Cavernous rock surfaces of the desert. *American Journal of Science, series 5*, 17, 393–399.

Blatt, H., Middleton, G. and Murray. R. (1972) *Origin of Sedimentary Rocks*. New Jersey: Prentice Hall.

Bonnell, D. G. R. and Nottage, M. E. (1939) Studies in porous material with special reference to building materials. I. The crystallisation of salts in porous materials. *Journal of the Society of Chemical Industry* 58A, 16–21.

Boss, G. (1941) Niederschlasmenge und Salzgehalt des Nebelswassers an der Kuste Deutsche-Sudwest Afrikas. *Bioklimatische Beiblätter der Meterologischen Zeitschrift* 8, 1–15.

Boulanger, C. and Urbain, G. (1912) Théorie de l'efflorescence des hydrates salins. *Comptes Rendus de l'Academie des Sciences* 155, 1246–1249.

Bowler, J. M. (1970) Late Quaternary environments; a study of lakes and associated sediments in south-eastern Australia. *Unpublished PhD Thesis*, Australian National University.

Bradley, W. C., Hutton, J. T. and Twidale, C. R. (1978) Role of salts in development of granitic tafoni, south Australia. *Journal of Geology* 86, 647–654.

Bromblet, P. (1993) Relations entre les variations des conditions environnementales et les processus de dégradation successive des temples de Karnak (Egypt). In: *Conservation of Stone and Other Materials* (Ed. M.-J. Thiel). London: Spon, 91–98.

Brown, P. W., Robbins, C. R. and Clifton, J. R. (1979) Adobe II: factors affecting the durability of adobe structures. *Studies in Conservation* 24, 23–39.

Bruckner, W. D. (1966) Salt weathering and inselbergs. *Nature*, 210, 832.

Bruiyn, H. de (1971) Pans in the Western Orange Free State. *Annals of the Geological Survey of South Africa* 9, 121–124.

Brunnacker, K. (1973) Einiges über löss-vorkommen in Tunisien. *Eiszeitaller Und Gegenwart* 23–24, 89–99.

Brunsden, D., Doornkamp, J. C. and Jones, D. K. C. (1979) The Bahrain Surface Materials Resources Survey and its application to regional planning. *Geographical Journal* 145, 1–35.

Buch, M. W. and Rose, D. (1996) Mineralogy and geochemistry of the sediments of the Etosha Pan Region in northern Namibia: a reconstruction of the depositional environment. *Journal of African Earth Sciences* 22(3), 355–378.

Bucher, A. and Lucas, G. (1975) Poussières africaines sur L'Europe. *La Météorologie* 33, 53–69.

Building Research Establishment (1986) Concrete in sulphate-bearing soils and groundwaters. *BRE Digest* 250.

Building Research Establishment (1987) Concrete, part 1: materials. *BRE Digest* 325.

Bulley, B. G. (1986) The engineering geology of Swakopmund. *Communications Geological Survey SW Africa/Namibia* 2, 7–12.

Busson, G. and Perthuisot, J. R. (1977) Interêt de la Sebkha el Malah (Sud Tunisien) pour l'interpretation des séries évaporitiques anciennes. *Sedimentary Geology* 19, 139–164.

Butler, P. R. and Mount, J. F. (1986) Corroded cobbles in southern Death Valley: their relationship to honeycomb weathering and lake shorelines. *Earth Surface Processes and Landforms* 11, 377–387.

Butlin, R. N., Cootes, A. T., Yates, T. J. S., Cooke, R. U. and Viles, H. A. (1988) A study of the degradation of building materials at Lincoln Cathedral, Lincoln, England. In: *Proceedings, 7th International Congress on the Deterioration and Conservation of Stone* (Eds J. D. Rodrigues, F. Henriques and F. Telmo Jeremias). Lisbon: Laboratorio Nacional de Enghenaria Civil, 246–255.

Butlin, R. N., Coote, A. T., Devenish, M., Hughes, I. S. C., Hutchens, C. M., Irwin, J. G., Lloyd, G. O., Massey, S. W., Webb, A. H. and Yates, T. J. S. (1992) Preliminary results from the analysis of stone tablets from the National Materials Exposure Programme (NMEP). *Atmospheric Environment* 26B, 189–198.

Cabrera, J. G. and Plowman, C. (1988) The mechanism and rate of attack of sodium sulphate on cement and cement/pfa pastes. *Advances in Cement Research* 1(3), 171–179.

Cailleux, A. (1968) Periglacial of the McMurdo Strait (Antarctica). *Biulteyn Peryglacjalny* 17, 57–116.

Cameron, R. E. (1969) Cold desert characteristics and problems relevant to other arid lands. In: *Arid Lands in Perspective* (Eds W. G. McGinnies and B. J. Goldman). Tuscon: University of Arizona Press, pp. 169–205.

Campbell, I. B. and Claridge, G. G. C. (1987) *Antarctica: Soils, Weathering, Processes and Environment.* Amsterdam: Elsevier.

Camuffo, D., Del Monte, M. and Sabbioni, C. (1983) Origin and growth mechanisms of the sulphated crusts on urban limestone. *Water, Air and Soil Pollution* 19, 351–359.

Carretero, M. I. and Galan, E. (1996) Marine spray and urban pollution as the main factors of stone damage in the Cathedral of Malaga (Spain). In: *Proceedings, 8th International Congress on Deterioration and Conservation of Stone* (Ed. J. Riederer). Berlin: Ernest and Sohn, 311–324.

Chapman, R. W. (1980) Salt weathering by sodium chloride in the Saudi Arabian Desert. *American Journal of Science* 280, 116–129.

Chappell, M. A. and Bartholomew, G. A. (1981). Standard operative temperatures and thermal energetics of the antelope ground squirrel, *Ammospermophitous leucurus. Physiological Zoology* 54, 81–93.

Charola, A. E. and Lewin, S. Z. (1979) Efflorescences on building stones — SEM in the characterization and elucidation of the mechanisms of formation. *Scanning Electron Microscopy* 1, 379–386.

Ciabach, J. (1996) The effect of water soluble salts on the impregnation of sandstone with silicone microemulsions. In: *Proceedings, 8th International Congress on Deterioration and Conservation of Stone* (Ed. J. Riederer). Berlin: Ernest and Sohn, 1215–1221.

Claridge, G. G. C. and Campbell, I. B. (1968) Origin of nitrate deposits. *Nature* 217, 428–430.

Claridge, G. G. C. and Campbell, I. B. (1977) The salts in Antarctic soils, their distribution and relationship to soil processes. *Soil Science* 123, 371–384.

Clarke, F. W. (1916) The data of geochemistry. *USGS Bulletin* 616.

Clarke, J. D. A. (1994) Geomorphology of the Kambalda region, Western Australia. *Australian Journal of Earth Sciences* **41**, 229–239.

Cloudsley-Thompson, J. L. and Chadwick, M. J. (1964) *Life in Deserts*. Foulis: London.

Coleman, J. M., Gagliano, S. M. and Smith, W. G. (1966) Chemical and physical weathering on saline high tidal flats, Northern Queensland, Australia. *Geological Society of America Bulletin* 77, 205–206.

Conca, J. L. and Astor, A. M. (1987) Capillary moisture flow and the origin of cavernous weathering in dolerites of Bull Pass, Antarctica. *Geology* 15, 151–154.

Conca, J. L. and Rossman, G. R. (1985) Core softening in cavernous weathered tonalite. *Journal of Geology* 93, 59–73.

Cooke, R. U. (1979) Laboratory simulation of salt weathering processes in arid environments. *Earth Surface Processes* 4, 347–359.

Cooke, R. U. (1994) Salt weathering and the urban water table in deserts. In: *Rock Weathering and Landform Evolution* (Eds D. A. Robinson and R. B. G. Williams). Chichester: Wiley, 193–205.

Cooke, R. U. and Gibbs, G. B. (no date) *Crumbling Heritage? Studies of Stone Weathering in Polluted Atmospheres*. London: National Power and Powergen.

Cooke, R. U. and Smalley, I. J. (1968) Salt weathering in deserts. *Nature* 220, 1226–1227.

Cooke, R. U., Brunsden, D., Doornkamp, J. C. and Jones, D. K. C. (1982) *Urban Geomorphology in Drylands*. Oxford: Oxford University Press.

Cooke, R. U., Warren, A. and Goudie, A. (1993) *Desert Geomorphology*. London: UCL Press.

Coque, R. (1961) *La Tunisie présaharienne, étude géomorphologique*. Paris: Colin.

Correns, C. W. (1949) Growth and dissolution of crystals under linear pressure. *Discussions of the Faraday Society, 5, Crystal Growth* 267–271.

Cotter, G. De P. (1923) The alkaline lakes and the soda industry of Sind. *Memoir Geological Survey of India* 47, 202–297.

Crammond, N. J. (1985) Thausamite in failed cement mortars and renders from exposed brickwork. *Cement and Concrete Research* 15, 1039–1050.

Custodio, E., Iribar, V., Manzano, B. A. and Galofre, A. (1986) Evolution of sea water chemistry in the Llobregat Delta, Barcelona, Spain. In: *Proceedings of the 9th Salt Water Chemistry Meeting, Delft*.

Dalongeville, R. (1995) Le rôle des organismes constructeurs dans la morphologie des littoraux de la Mer Méditerranée: algues calcaires et vermetidés. *Norois* 165, 73–88.

Danin, A. and Garty, J. (1983) Distribution of cyanobacteria and lichens on hillsides of the Negev Highlands and their impact on biogenic weathering. *Zeitschrift für Geomorphologie* 27, 423–444.

De Decker, P. (1983) Australian salt lakes: their history, chemistry and biota — a review. *Hydrobiologia* 105, 231–244.

De Witte, E. and Bos, K. (1992) Conservation of ferruginous sandstone used in northern Belgium. In: *Proceedings, 7th International Congress on the Deterioration and Conservation of Stone* (Eds J. D. Rodrigues, F. Henriques and F. Telmo Jeremias). Lisbon: Laboratorio Nacional de Enghenaria Civil, 1113–1121.

De Witte, E. and Dupas, M. (1992) Cleaning poultices based on E.D.T.A. In: *Proceedings, 7th International Congress on Deterioration and Conservation of Stone* (Eds J. D. Rodrigues, F. Henriques and F. Telmo Jeremias). Lisbon: Laboratorio Nacional de Enghenaria Civil, 1023–1028.

Debenham, F. (1920) A new mode of transportation by ice: the raised marine muds of south Victoria Land (Antarctica). *Quarterly Journal of the Geological Society of London* 75, 51–76.

del Monte, M. (1991) Trajan's Column: lichens don't live here anymore. *Endeavour* 15, 86–93.

del Monte, M. and Sabbioni, C. (1984) Gypsum crusts and fly ash particles on carbonatic outcrops. *Archives for Meteorology, Geophysics and Bioclimatology* Series B 35, 105–111.

del Monte, M. and Vittori, O. (1985) Air pollution and stone decay: the case of Venice. *Endeavour NS* 9(3), 117–122.

Dobson, M. C. (1991) De-icing salt damage to trees and shrubs. *Forestry Commission Bulletin* 101.

Doornkamp, J. C. and Ibrahim, H. A. M. (1990) Salt weathering. *Progress in Physical Geography* 14, 335–348.

Doornkamp, J. C., Brunsden, D. and Jones, D. K. C. (1980) *Geology, Geomorphology and Pedology of Bahrain*. Norwich: Geobooks.

Dort, W. and Dort, D. S. (1970a) Low temperature origin of sodium sulphate deposits: particularly in Antarctica. In: *Third Symposium on Salt* (Eds J. L. Rau and L. F. Dellwig). 181–205.

Dort, W. and Dort, D. S. (1970b) Sodium sulfate deposits in Antarctica. *Modern Geology* 1, 91–117.

Dow, J. A. S. and Neall, V. E. (1974) Geology of the Lower Rennick glacier, northern Victoria Land, Antarctica. *New Zealand Journal of Geology and Geophysics* 17, 659–714.

Dragovich, D. (1969) The origin of cavernous surfaces (tafoni) in granitic rocks of southern south Australia. *Zeitschrift für Geomorphologie* 13, 163–181.

Dragovich, D. (1994) Fire, climate and the persistence of desert varnish near Dampier, Western Australia. *Palaeogeography, Palaeoclimatology, Palaeoecology* 111, 279–288.

Dreissen, P. M. and Schoorl, J. (1973) Mineralogy and morphology of salt efflorescences on saline soils in the Great Konya Basin, Turkey. *Journal of Soil Science* 24, 436–424.

Du Toit, A. L. (1906) Geological survey of portions of the divisions of Vryburg and Mafeking. In: *10th Annual Report Geological Commission of The Cape of Good Hope*. 205–258.

Ducloux, J., Guero, Y., Fallavier, P. and Valet, S. (1994) Mineralogy of salt efflorescences in paddy field soils of Kollo, Southern Niger. *Geoderma* 64, 57–71.

Duffy, A. P. and Perry, S. H. (1996) The mechanisms and causes of Portland limestone decay—a case study. In: *Proceedings, 8th International Congress on the Deterioration and Conservation of Stone* (Ed. J. Riederer). Berlin: Ernst und Sohn, 135–145.

Dunn, E. J. (1915) Geological notes, Northern Territory, Australia. *Proceedings, Royal Society of Victoria* 28 (NS), 112–114.

Egorov, A. N. (1993) Mongolian salt lakes: some features of their geography, thermal patterns, chemistry and biology. *Hydrobiologia* 267, 13–21.

Eriksen, G. E. (1981) Geology and origin of the Chilean nitrate deposits. *USGS Professional Paper* 1188.

Eriksson, E. (1958) The chemical climate and saline soils in the Arid Zone. *Arid Zone Research, UNESCO* 10, 147–180.

Eriksson, E. (1960) The yearly circulation of chloride and sulfur in nature: meteorological and pedological implications, Part II. *Tellus* 12, 63–109.

Eugster, H. P. and Jones, B. F. (1979) Behaviour of major solutes during closed-basin brine evolution. *American Journal of Science* 279, 609–613.

Eugster, H. P. and Maglione, G. (1979) Brines and evaporites of the Lake Chad basin, Africa. *Geochimica et Cosmochimica Acta* 43, 973–981.

Eugster, H. P. and Smith, G. I. (1965) Mineral equilibria in the Searles Lake evaporites California. *Journal of Petrology* 6, 473–522.

Evans, G. (1995) The Arabian Gulf: a modern carbonate–evaporite factory; a review. *Cuadernos de Geología Ibérica* **19**, 61–96.

Evans, I. S. (1970) Salt crystallisation and rock weathering: a review. *Revue de Géomorphologie Dynamique* **19**, 153–177.

Everett, D. H. (1961) The thermodynamics of frost damage to porous solids. *Transactions, Faraday Society* **57**, 1541–1551.

Fahey, B. (1985) Salt weathering as a mechanism of rock breakup in cold climates: an experimental approach. *Zeitschrift für Geomorphologie* **29**, 99–111.

Fassina, V. (1988) The stone decay of the main portal of Saint Mark's Basilica in relation to natural weathering agents and to air pollution. In: *Proceedings, 6th International Congress on the Deterioration and Conservation of Stone*. Torun: Nicolas Copernicus University Press Department, 276–286.

Fassina, V., Arbizzani, R. and Naccari, A. (1996) Salt efflorescence on the marble slabs of S. Maria dei Miracoli church: a survey on their origin and on the methodology of their removal. In: *Proceedings, 8th International Congress on Deterioration and Conservation of Stone*, (Ed. J. Reiderer) Berlin: Ernst und Sohn, 523–534.

Fitzner, B. (1994) Porosity properties and weathering behaviour of natural stones — methodology and examples. In: *Stone Materials in Monuments: Diagnosis and Conservation* (Ed. F. Zezza). C.U.M. University School of Monument Conservation, Second Course, Crete, May 1993, 43–54.

Fitzner, B. and Heinrichs, K. (1991) Weathering forms and rock characteristics of historical monuments carved from bedrock in Petra/Jordan. In: *Science, Technology and European Cultural Heritage* (Eds N. S. Baer, C. Sabbioni and A. I. Sors). Oxford: Butterworth-Heinemann, 908–911.

Fitzner, B. and Heinrichs, K. (1994) Damage diagnosis at monuments carved from bedrock in Petra, Jordan. In: *Proceedings, Third International Conference on the Conservation of Monuments in the Mediterranean Basin, Venice, 22–25 June, 1994* (Eds V. Fassina, H. Ott and F. Zezza). Venice: Soprintendenza ai Beni Artistica e Storici di Venezia, 663–671.

Fitzner, B. and Snethlage, R. (1982) Einfluss der Porenradienverteilung auf das Verwitterungsverhalten ausgewählter Sandsteine. *Bautenschutz und Bausanierung* **5**, 97–103.

Fitzner, B., Heinrichs, K. and Kownatzki, R. (1995) Weathering forms — classification and mapping. In: *Fortdruck ans Denkmalpflege und Naturwissenschaft Natursteinkonservierung* 1 (Ed. R. Snethlage). Berlin: Ernst und Sohn, 41–88 (in German and English).

Focke, J. W. (1978) Limestone cliff morphology on Curaçao (Netherlands Antilles), with special attention to the origin of notches and vermetid/coralline algal surf benches ('cornices', 'trottoirs'). *Zeitschrift für Geomorphologie* **22**, 329–349.

Fookes, P. G. and French, W. J. (1977) Soluble salt damage to surfaced roads in the Middle East. *The Highway Engineer* **24** (Dec), 10–20.

Fookes, P. G. and Knill, J. L. (1969) The application of engineering geology in the regional development of northern and central Iran. *Engineering Geology* **3**, 81–120.

Ford, D. C. and Williams, P. A. (1989) *Karst Geomorphology and Hydrology*. London: Unwin Hyman.

Foster, W. R. and Hoover, K. V. (1963) Hexahydrate ($MgSO_4 \cdot 6H_2O$) as an efflorescence on some Ohio dolomites. *Ohio Journal of Science* **63**, 158.

Friedman, E. J. (1980) Endolithic microbial life in hot and cold deserts. *Origins of Life* **10**, 227–235.

Galan, E. (1990) Carbonate rocks; alteration and control of stone quality: some considerations. In: *The Conservation of Monuments in the Mediterranean Basin* (Ed. F. Zezza). Brescia: Grafo, 249–254.

Garcia-Vallès, M., Mohera, J. and Vendrell-Saz, M. (1996) Man induced deterioration in the Gothic Clorister of "Palau de la Generalitat" (Barcelona–Spain). In: *Proceedings, 8th International Congress on Deterioration and Conservation of Stone* (Ed. J. Riederer). Berlin: Ernst und Sohn, 61–70.

Gavish, E. (1980) Recent sabkhas marginal to the southern coasts of Sinai, Red Sea. *Developments in Sedimentology* **28**, 233–251.

Ghassemi, F., Jakeman, A. J. and Nix, H. A. (1995) *Salinisation of Land and Water Resources*. Wallingford: CAB International, 536 pp.

Gibson, G. W. (1962) Geological investigations in Southern Victoria Land, Antarctica. *New Zealand Journal of Geology and Geophysics* **5**, 361–374.

Gillieson, D. (1996) *Caves: Processes, Development and Management*. Oxford: Blackwell.

Gillott, J. E. (1979) Effect of deicing agents and sulphate solutions on concrete aggregate. *Quarterly Journal of Engineering Geology* **11**, 177–192.

Gindy, A. R. (1991) Orgin of the Qattara Depression, Egypt. Discussion. *Bulletin Geological Society of America* **103**, 1374–1376.

Gindy, A. R. and El Askary, M. A. (1969) Stratigraphy, structure and origin of Siwa Depression, Western Desert of Egypt. *Bulletin of the American Association of Petroleum Geologists* **53**, 603–625.

Gleick, P. H. (Ed.) (1993) *Water in Crisis: a Guide to the World's Fresh Water Resources*. New York and Oxford: Oxford University Press.

Godbole, N. N. (1972) Theories on the origin of salt lakes in Rajasthan, India. In: *24th International Geological Congress, Section 10*, 345–347.

Götürk, H., Volkan, M. and Kahveci, S. (1993) Sulfation mechanism of travertines: effect of SO_2 concentration, relative humidity and temperature. In: *Conservation of Stone and Other Materials* (Ed. M.-J. Thiel). London: Spon, 83–90.

Goudharzi, G. H. (1970) *Nonmetallic Mineral Resources: Saline Deposits, Silica Sand, Sulfur and Trona*. USGS Professional Paper **660**, 82–93.

Goudie, A. S. (1973) *Duricrusts of Tropical and Sub-tropical Landscapes*. Oxford: Clarendon Press.

Goudie, A. S. (1974) Further experimental investigation of rock weathering by salt and other mechanical processes. *Zeitschrift für Geomorphologie Supplementband* **21**, 1–12.

Goudie, A. S. (1977) Sodium sulphate weathering and the disintegration of Mohenjo-Daro, Pakistan. *Earth Surface Processes* **2**, 75–86.

Goudie, A. S. (1978) Dust storms and their geomorphological implications. *Journal of Arid Environments* **1**, 291–310.

Goudie, A. S. (1984) Salt efflorescences and salt weathering in the Hunza Valley, Karakoram Mountains, Pakistan. In: *Proceedings, International Karakoram Project* (Ed. K. J. Miller). Cambridge: University Press, 607–615.

Goudie, A. S. (1986) Laboratory simulation of 'the wick effect' in salt weathering of rock. *Earth Surface Processes and Landforms* **11**, 275–285.

Goudie, A. S. (1993) Salt weathering simulation using a single-immersion technique. *Earth Surface Processes and Landforms* **18**, 369–376.

Goudie, A. S. and Cooke, R. U. (1984). Salt efflorescences and saline lakes; a distributional analysis. *Geoforum* **15**, 563–582.

Goudie, A. S. and Day, M. (1980) Disintegration of fan sediments in Death Valley, California, by salt weathering. *Physical Geography* **1**, 126–137.

Goudie, A. S. and Thomas, D. S. G. (1985) Pans in southern Africa with particular reference to South Africa and Zimbabwe. *Zeitschrift für Geomorphologie* **29**, 1–19.

Goudie, A. S. and Viles, H. A. (1995) The nature and pattern of debris liberation by salt weathering: a laboratory study. *Earth Surface Processes and Landforms* **20**, 437–449.

Goudie, A. S. and Watson, A. (1984) Rock block monitoring of rapid salt weathering in southern Tunisia. *Earth Surface Processes and Landforms* **9**, 95–99.

Goudie, A. S. and Wells, G. L. (1995) The nature, distribution and formation of pans in arid zones. *Earth-Science Review* **38**, 1–69.

Goudie, A. S., Cooke, R. U. and Evans, I. S. (1970) Experimental investigation of rock weathering by salts. *Area* **4**, 42–48.

Goudie, A. S., Cooke, R. U. and Doornkamp, J. C. (1979) The formation of silt from quartz dune sand by salt-weathering processes in deserts. *Journal of Arid Environments* **2**, 105–112.

Goudie, A. S., Jones, D. K. C. and Brunsden, D. (1984) Recent fluctuations in some glaciers of the western Karakoram Mountains, Hunza, Pakistan. In: *Proceedings of the International Karakoram Project* (Ed. K. J. Miller). Cambridge: Cambridge University Press, 411–455.

Goudie, A. S., Viles, H. A. and Parker, A. G. (in press) Monitoring of rapid limestone decay in the Central Namib Desert by the rock block method. *Journal of Arid Environments*.

Guilcher, A. and Bodéré, J. C. (1975) Formes de corrosion littorale dans les roches volcaniques aux moyennes et hautes latitudes dans l'Atlantique. *Bulletin Association Geographes Française* **426**, 179–185.

Gumuzzio, J., Battle, J. and Casas, J. (1982) Mineralogy composition of salt efflorescences in a typic salorthid, Spain. *Geoderma* **28**, 39–51.

Gundel, L. A., Benner, W. H. and Hansen, A. D. A. (1994) Chemical composition of fogwater and interstitial aerosol in Berkeley, California. *Atmospheric Environment* **28**, 2715–2725.

Gustafsson, M. E. R. and Franzen, L. G. (1996) Dry deposition and concentration of marine aerosols in a coastal area, S W Sweden. *Atmospheric Environment* **30**, 977–989.

Hall, K. J. and Walton, D. W. H. (1992) Rock weathering, soil development and colonization under a changing climate. *Philosophical Transactions Royal Society of London, B* **338**, 269–277.

Hammecker, C. and Jeannette, D. (1988) Role des propriètes physiques dans l'altération de roches carbonatées: exemple de la façade ouest de Notre-Dame-La Grande de Poitiers (France) In: *Proceedings, 6th International Congress on the Deterioration and Conservation of Stone.* Torun: Nicolas Copernicus University Press Department, 266–275.

Haneef, S. J., Johnson, J. B., Dickinson, C., Thompson, G. E. and Wood, G. C. (1992) Effect of dry deposition of NO_x and SO_2 gaseous pollutants on the degradation of calcareous building stones. *Atmospheric Environment* **26A**, 2963–2974.

Hansen, K. (1970) Geological and geographical investigations in Kong Frederik Ixs land. *Meddelelser am Grønland* **188** (4).

Hardie, L. A. (1968) The origin of the recent non-marine evaporite deposit of Saline Valley, Inyo County, California. *Geochimica et Cosmochimica Acta* **32**, 1279–1301.

Harris, H. J. H., Cartwright, K. and Troll, T. (1979) Dynamic chemical equilibrium in a polar desert pond: a sensitive index of meteorological cycles. *Science* **204**, 301–303.

Hawass, Z. (1993) The Egyptian monuments: problems and solutions. In: *Conservation of Stone and Other Materials* (Ed. M.-J. Thiel). London: Spon, 19–25.

Hawkins, A. B. and Pinches, G. M. (1987) Cause and significance of heave at Llandrough Hospital, Cardiff — a case history of ground floor heave due to crystal growth. *Quarterly Journal of Engineering Geology* **20**, 41–57.

Hayden, J. D. (1945) Salt erosion. *American Antiquity* **10**, 373–378.

Haynes, C. V. (1982) The Darb El-Arba'in desert: a product of Quaternary climate change. In: *Desert Landforms of Southwest Egypt: a Basis for Comparsion with Mars* (Eds F. El-Baz and T. A. Maxwell). Washington: NASA, 91–117.

Heathcote, R. L. (1983) *The arid lands: their use and abuse*. London: Longman.

Helmi, F. M. (1990) Study of salt problem in the Sphinx, Giza, Egypt. In: *ICCOM Committee for Conservation 9th Triennial Meeting, Dresden, 26–31 August 1990, Preprints 1*, 326–329.

Herodotus (1972) *The Histories* (translated by A. De Sélincourt). Harmondsworth: Penguin Books.

Herut, B., Spiro, B., Starinsky, A. and Ktaz, A. (1995) Sources of sulphur in rainwater as indicated by isotopic $\delta^{34}S$ data and chemical composition, Israel. *Atmospheric Environment* **29**, 851–857.

Hewitt, K. (1969) Geomorphology of the mountainous regions of the upper Indus basin. Unpublished PhD Dissertation, University of London, 2 vols, 310 + 121 pp.

Hobbs, W. H. (1917) The erosional degradational processes of deserts, with special reference to the origin of desert depressions. *Annals of the Association of American Geographers* **7**, 25–60.

Höllerman, P. (1975) Formen Kavernöser Verwitterung ("Tafoni") Auf Teneriffa. *Catena* **2**, 385–410.

Hong Naifeng (1994) Corrosiveness of the arid saline soil in China. In: *Engineering Characteristics of Arid Soils* (Eds P. Fookes and R. H. G. Parry). Rotterdam: Balkema, 29–34.

Horta, J. C. de O. S. (1985) Salt heaving in the Sahara. *Géotechnique* **35**, 329–337.

Howard, K. W. F. and Beck, P. J. (1993) Hydrogeochemical implications of groundwater contamination by road de-icing chemicals. *Journal of Contaminant Hydrology* **12**, 245–268.

Hudec, P. P. and Rigbey, S. G. (1976) The effect of sodium chloride on water sorption characteristics or rock aggregate. *Bulletin of the Association of Engineering Geologists* **13**, 199–211.

Hugo, P. J. (1974) Salt in the Republic of South Africa. *Memoir Geological Survey of South Africa* **65**, 105 pp.

Hume, W. F. (1925) *Geology of Egypt 1*. Cairo: Government Press.

Hunt, C. B. (1975) *Death Valley, Geology, Ecology, Archaeology*. Berkeley: University of California Press.

Hunt, C. B., Robinson, T. W., Bowles, W. A. and Washburn, A. L. (1966) *Hydrologic Basin Death Valley, California*. USGS Professional Paper **494–4**.

Hutchinson, A. J., Johnson, J. B., Thompson, G. E. and Wood, G. C. (1992) Stone degradation due to dry deposition of HCl and SO_2 in a laboratory-based exposure chamber. *Atmospheric Environment* **26A**, 2785–2793.

Ibrahim, H. A. M. and Doornkamp, J. C. (1991) Towns of the Nile Delta and the potential for damage from aggressive saline groundwater. *Third World Planning Review* **13**, 83–90.

Imeson, A. and Emmer, I. M. (1992) Implications of climatic change on land degradation in the Mediterranean. In: *Climatic Change and the Mediterranean* (Eds L. Jeftic, J. D. Milliman and G. Sestini). London: Arnold, 95–128.

Jansen, M. (1996) Save Mohenjo-Daro? In: *Forgotten Cities on the Indus* (Eds M. Jansen, M. Mulloy and G. Urban). Mainz: Verlag Philipp von Zabern, 220–234.

Januszke, R. M. and Booth, E. H. S. (1984) Soluble salt damage to sprayed seals on the Stuart Highway. *Australian Road Research Board Proceedings* **12**(3), 18–31.

Jarman, M. (1994) Costing the benefits of the Oxford Transport Strategy. *Transportation Planning Systems* **2**, 23–37.

Jefferson, D. P. (1993) Building stone: the geological dimension. *Quarterly Journal of Engineering Geology* **26**, 305–319.

Jenkins, W. J. (1934) *Annual Report of the Department of Agriculture* in Sind, Bombay.

Jennings, J. N. (1985) *Karst Geomorphology*. Oxford: Blackwell.

Jerwood, L. C., Robinson, D. A. and Williams, R. B. G. (1990a) Experimental frost and salt weathering of chalk — I. *Earth Surface Processes and Landforms* 15, 611–624.

Jerwood, L. C., Robinson, D. A. and Williams, R. B. G. (1990b) Experimental frost and salt weathering of chalk — II. *Earth Surface Processes and Landforms* 15, 699–708.

Johannessen, C. L., Feiereisen, J. J. and Wells, A. K. (1982) Weathering of ocean cliffs by salt expansion in mid-latitude coastal environment. *Shore and Beach* 50, 26–34.

Johannesson, J. K. and Gibson, G. W. (1962) Nitrate and iodate in Antarctica salt deposits. *Nature* 194, 567–568.

Johnson, M. (1980) The origin of Australia's salt lakes. *Records, New South Wales Geological Survey* 19, 221–266.

Johnston, J. H. (1973) Salt weathering processes in the McMurdo dry valley regions of South Victoria Land, Antarctica. *New Zealand Journal of Geology and Geophysics* 16, 221–224.

Jones, B. F. (1965) *The Hydrology and Mineralogy of Deep Springs Lake, Inyo County, California*. USGS Professional Paper 502-A.

Jones, B. F., Hanor, J. S. and Evans, W. R. (1994) Sources of dissolved salts in the central Murray Basin, Australia. *Chemical Geology* 111, 125–154.

Jones, D. K. C. (1980) British applied geomorphology: an appraisal. *Zeitschrift für Geomorphologie. N. F., Suppl.-Bd* 36, 48–73.

Jones, L. M., Faure, G., Taylor, K. S. and Corbato, C. E. (1983) The origin of salts on Mount Erebus and along the coast of Ross Island, Antarctica. *Isotope Geoscience* 1, 57–64.

Jones, M. S., O'Brien, P. F., Haneef, S. J., Thompson, G. E., Wood, G. C. and Cooper, T. P. (1996) A study of decay occurring in Leinster granite, House No. 9, Trinity College, Dublin. In: *Proceedings, 8th International Congress on Deterioration and Conservation of Stone* (Ed. J. Reiderer), Berlin: Ernst und Sohn, 211–221.

Jutson, J. T. (1918) The influence of salts in rock weathering in sub-arid Western Australia. *Proceedings, Royal Society of Victoria* 80, 165–172.

Jutson, J. T. (1950) *The Physiography (Geomorphology) of Western Australia* (3rd edn). Perth: Government Printer.

Kayyali, O. A. (1989) Porosity and compressive strength of cement paste in sulphate solution. *Cement and Concrete Research* 19, 423–433.

Keller, L. P., McCarthy, G. J. and Richardson, J. L. (1986) Laboratory modeling of northern Great Plains salt efflorescence mineralogy. *Soil Science Society of America Journal* 50, 1363–1367.

Kelletat, D. (1988) Quantitative investigations on coastal bioerosion in higher latitudes: an example from northern Scotland. *Geoökodynamik* 9, 41–51.

Keunen, P. H. (1969) Origin of quartz silt. *Journal of Sedimentary Petrology* 39, 1631–1633.

Keys, J. R. and Williams, K. (1981) Origin of crystalline, cold desert salts in the McMurdo Region, Antarctica. *Geochimica et Cosmochimica Acta* 45, 2299–2309.

Kirchner, G. (1995) Physikalische Verwitterung in Trockengebieken unter Betonung der Salzverwitterung am Beispiel des Basin-and-Range Gebiets (südwestliche USA und nördlisches Mexico). *Mainzer Geographische Studien* 41, 267 pp.

Kirchner, G. (1996) Cavernous weathering in the Basin and Range area, southwestern USA and northern Mexico. *Zeitschrift für Geomorphologie Supplementband* 106, 73–97.

Kirkitsos, P. and Sikiotis, D. (1995) Deterioration of Pentelic marble, Portland limestone and Baumberger sandstone in laboratory exposures to gaseous nitric acid. *Atmospheric Environment* 29, 77–86.

Klaer, W. (1993) Beobachtungen zu Formen der Verwitterung und Abtragung auf Sansteinfelsen und Graniten in der algerischen Sahara. *Würzburger Geographische Arbeiten* 87, 121–132.

Klemm, W. and Siedel, H. (1996) Sources of sulphate salt efflorescences at historical monuments — a geochemical study from Freiberg, Saxony. In: *Proceedings, 8th International Congress on Deterioration and Conservation of Stone* (Ed. J. Riederer). Berlin: Ernst und Sohn, 489–495.

Knacke, O. and Erdberg R. (1975) The crystallisation pressure of sodium sulphate decahydrate. *Berichte der Bunsen-Gesellschaft für Physikalische Chemice* 79, 653–657.

Knöfel, D. K., Hoffmann, D. and Snethlage, R. (1987) Physico-chemical weathering reactions as a formulary for time-lapsing ageing tests. *Materials and Structures* 20, 127–145.

Koestler, R. J., Charola, A. E. and Wypyski, M. (1985) Microbiologically induced deterioration of dolomitic and calcitic stone as viewed by scanning electron microscopy. In: *Proceedings, 5th International Congress on Deterioration and Conservation of Stone* (Ed. G. Felix). Lausanne: Presses Polytechniques Romandes, 617–626.

Kohut, C. K. and Dudas, M. J. (1993) Evaporite mineralogy and trace-element content of salt-affected soils in Alberta. *Canadian Journal of Soil Science* 73, 399–409.

Kolm, K. E. (1982) Predicting the surface wind characteristics of southern Wyoming from remote sensing and aeolian geomorphology. *Geological Society of America Special Paper* 192, 25–53.

Kouzeli, K. (1992) Black crust removal methods in use. Their effects on Pentelic marble surfaces. In: *Proceedings, 7th International Congress on Deterioration and Conservation of Stone* (Eds J. D. Rodrigues, F. Henriques and F. Telmo Jeremias). Lisbon: Laboratorio Nacional de Enghenaria Civil, 1147–1156.

Kozlowski, R., Magiera, J., Weber, J. and Haber, J. (1990) Decay and conservation of Pinczów limestone I. Lithology and weathering. *Studies in Conservation* 35, 205–221.

Kozlowski, R., Tokarz, M. and Persson, M. (1992a) "Gypstop" — a novel protective treatment. In: *Proceedings, 7th International Congress on Deterioration and Conservation of Stone* (Eds J. D. Rodrigues, F. Henriques and F. Telmo Jeremias). Lisbon: Laboratorio Nacional de Enghenaria Civil, 1187–1196.

Kozlowski, R., Hejda, A., Ceckiewicz, S. and Haber, J. (1992b) Influence of water contained in porous limestone on corrosion. *Atmospheric Environment* 26A, 1241–1248.

Kukal, Z. and Saadallah, A. (1970) Composition and rate of deposition of dust storm sediments in Central Iraq. *Casopsi mineralogii a geologii* 15, 227–234.

Kumar, S. and Kameswara Rao, C. V. S. (1994) Effect of sulphates on the setting time of cement and strength of concrete. *Cement and Concrete Research* 24, 1237–1244.

Kwaad, F. J. M. (1970) Experiments on the disintegration of granite by salt action. *University Amsterdam Fysisch Geografisch en Bodemkundig Laboratorium Publicatie* 16, 67–80.

La Iglesia, A., Garcia del Cura, M. A. and Ordoñez, S. (1994) The physicochemical weathering of monumental dolostones, granites and limestones; dimension stones of the Cathedral of Toledo (Spain). *The Science of the Total Environment* 152, 179–188.

Lageat, Y. (1994) La désert du Namib central. *Annales de Géographie* 578, 339–360.

Laity, J. E. and Malin, M. C. (1985) Sapping processes and the development of theater-headed valley networks on the Colorado Plateau. *Bulletin of the Geological Society of America* 96, 203–217.

Lal Gauri, K., Chowdhury, A. N., Kulshreshtha, N. P. and Punuru, A. R. (1989) The sulfation of marble and the treatment of gypsum crusts. *Studies in Conservation* 34, 201–206.

Lancaster, J., Lancaster, N. and Sealy, M. K. (1984) Climate of the central Namib Desert. *Madoqua* 14, 5–61.

Larsen, H. (1980) Ecology of hypersaline environments. *Developments in Sedimentology* **28**, 23–39.

Larson, J. H. (1992) New approaches to the conservation of external stone sculpture: the twelfth century frieze at Lincoln cathedral. In: *Proceedings, 7th International Congress on Deterioration and Conservation of Stone, Lisbon* (Eds J. Delgado Rodrigues, F. L. Henriques and F. Telmo Jeremias), 1167–1176.

Last, W. M. and Schweyen, T. H. (1983) Sedimentology and geochemistry of saline lakes of the Great Plains. *Hydrobiologia* **105**, 245–263.

Laue, S., Bläuer Böhm, C. and Jeannette, D. (1996) Salt weathering and porosity — examples from the crypt of St. Maria in Kapitol, Cologne. In: *Proceedings, 8th International Congress on Deterioration and Conservation of Stone* (Ed. J. Riederer). Berlin: Ernst und Sohn, 513–522.

Laurie, A. P. (1925) Stone decay and the preservation of buildings. *Journal of the Society of Chemical Industries* **44B**, 86–92.

Laurie, A. P. and Milne, J. (1926) The evaporation of water and salt solutions from surfaces of stone, brick and mortar. *Proceedings of the Royal Society of Edinburgh* **47**, 52–68.

Levy, Y. (1977) The origin and evolution of brine in coastal sabkha, northern Sinai. *Journal of Sedimentary Petrology* **47**, 451–462.

Levy, Y. (1980) Evaporitic environments in northern Sinai (Ed. A. Nissenbaum). *Developments in Sedimentology* **28**, 131–143.

Lewin, S. Z. (1981) The mechanism of masonry decay through crystallization. In: *Proceedings of a Conference on Conservation of Historic Stone Buildings and Monuments*. Washington: National Academy of Sciences, 120–144.

Lewin, S. Z. (1990) The susceptibility of calcareous stones to salt decay. In: *The Conservation of Monuments in the Mediterranean Basin* (Ed. F. Zezza). Brescia: Grafo, 59–63.

Lines, G. C. (1979) *Hydrology and Surface Morphology of the Bonneville Salt Flats and Pilot Valley Playa, Utah.* USGS Water Supply Paper **2057**.

Littmann, K., Sasse, H. R., Wagener, S. and Hocker, H. (1993) Development of polymers for the consolidation of natural stone. In: *Conservation of Stone and Other Materials* (Ed. M.-J. Thiel). London: Spon, 689–696.

Litvan, G. G. (1972) Phase transitions of adsorbates: IV, mechanism of frost action in hardened cement paste. *Journal of the American Ceramic Society* **55**, 38–42.

Litvan, G. G. (1975) Phase transitions of adsorbates: VI, effect of deicing agents on the freezing of cement paste. *Journal of the American Ceramic Society* **58**, 26–30.

Litvan, G. G. (1976) Frost action in cement in the presence of de-icers. *Cement and Concrete Research* **6**, 351–356.

Livingston, R. A. (1989) The geological origins of the Great Sphinx and implications for its preservation. *Abstract of paper presented to International Geological Congress, Washington DC, July 9–19, 1989.*

Livingston, R. A. (1994) Influence of evaporite minerals on gypsum crusts and alveolar weathering. In: *Proceeedings, IIIrd International Symposium on the Conservation of Monuments in the Mediterranean Basin, Venice.* (Eds V. Fassina, H. Ott and F. Zezza). 101–107.

Logan, J. (1974) African dusts as a source of solutes in Gran Canarian ground waters. *Geological Society of America, Abstracts of Program* **6**, 849.

Lohuizen-DeLeeuw, J. E. van (1973) Moenjo Daro — a cause of common concern. *South Asian Archaeology 1973.* Leiden: Brill, 1974, 1–11.

Lucas, A. (1915) *The disintegration of Building Stone.* Cairo: Government Printer.

Lukas, W. (1975) Betonzer störung durch SO_3 — Angriff uter bildung von Thaumasit und Woodfordit. *Cement and Concrete Research* **5**, 503–518.

Luquer, L. McJ. (1895) The relative effects of frost and the sulphate of soda efflorescence tests on building stones. *Transactions, American Society of Civil Engineers* 33, 235–256.

Mainguet, M. (1995) *L'homme et la sécheresse*. Paris: Masson.

Malin, M. C. (1974) Salt weathering on Mars. *Journal of Geophysical Research* 79, 3889–3894.

Mallett, G. P. (1995) *Repair of Concrete Bridges*. London: Thomas Telford.

Maravelaki, P., Biscontin, G., Polloni, R., Cecchetti, W. and Zendri, E. (1992) Investigation on surface alteration of limestone related to cleaning processes. In: *Proceedings, 7th International Congress on Deterioration and Conservation of Stone* (Eds J. D. Rodrigues, F. Henriques and F. Telmo Jeremias). Lisbon: Laboratorio Nacional de Enghenaria Civil, 1093–1101.

Marshall, J. (1973) *Mohenjo-Daro and the Indus Civilization*. Delhi: Indological Book House.

Martin, H. (1963) A suggested theory for the origin and a brief description of some gypsum deposits of South-West Africa. *Transactions of the Geological Society of South Africa* 66, 345–351.

Martin, L., Bellow, M. A. and Martin, A. (1992) The efflorescences of the Cathedral of Almeria (Spain). In: *Proceedings, 7th International Congress on the Deterioration and Conservation of Stone* (Eds J. D. Rodrigues, F. Henriques and F. Telmo Jeremias). Lisbon: Laboratorio Nacional de Enghenaria Civil, 869–873.

Martini, I. P. (1978) Tafoni weathering with examples from Tuscany, Italy. *Zeitschrift für Geomorphologie* 22, 44–67.

Martin-Penela, A. J. (1994) Pipe and gully systems development in the Alamanzora Basin (south east Spain) *Zeitschrift für Geomorphologie* 38, 207–222.

Master Plan (1972) *Master Plan for the Preservation of Moenjodaro*. Department of Archaeology and Museums, Ministry of Education and Provincial Coordination, Government of Pakistan.

Matsukura, Y. and Kanai, H. (1988) Salt fretting in the Valley Cliff of the Asama Volcano Region, Japan. *Earth Surface Processes and Landforms* 13, 85–90.

Matsukura, Y. and Matsuoka, N. (1991) Rates of tafoni weathering on uplifted shore platforms in Nojima-Zaki, Boso Penisula, Japan. *Earth Surface Processes and Landforms* 16, 51–56.

Matsukura, Y. and Yatsu, E. (1985) Influence of salt water on slaking rate of Tertiary shale and tuff. *Transactions, Japanese Geomorphological Union* 6–2, 163–167.

Matsuoka, N. (1995) Rock weathering processes and landform development in the Sør Rondane Mountains, Antarctica. *Geomorphology* 12, 323–339.

Matsuoka, N., Motiuaki, K. and Kirakawa, K. (1996) Field experiments on physical weathering and wind erosion in an Antarctic cold desert. *Earth Surface Processes and Landforms* 21, 687–699.

Mazor, E. and Mantel, M. (1966) Epsomite efflorescence and the composition of shallow ground waters in the southern Negev, Israel. *Israel Journal of Earth-Sciences* 15, 71–75.

McCardle, N. C. and Liss, P. S. (1995) Isotopes and atmospheric sulphur. *Atmospheric Environment* 29, 2553–2556.

McGee, E. S. and Mossotti, V. G. (1992) Gypsum accumulation on carbonate stone. *Atmospheric Environment* 26B, 249–253.

McGreevy, J. P. (1982) 'Frost and salt' weathering: further experimental results. *Earth Surface Processes and Landforms* 7, 475–488.

McGreevy, J. P. (1985) A preliminary scanning electron microscope study of honeycomb weathering of sandstone in a coastal environment. *Earth Surface Processes and Landforms* 10, 509–518.

McInnis, C. and Whiting, J. D. (1979) The frost resistance of concrete subjected to a deicing agent. *Cement and Concrete Research* 9, 325–336.

McLeod, I. R. (1964) Saline lakes of the Vestfold Hills, Princess Elizabeth Land. In: *Antarctic Geology* (Ed. R. J. Adie). Amsterdam: North Holland, 65–72.

McNally, G. H. (1995) Engineering characteristics and uses of duricrusts in Australia. *Australian Journal of Earth Sciences* 42, 535–547.

McNamara, E. E. and Usselman, T. (1972) Salt minerals in soils profiles and as surficial crusts and efflorescences, coastal Enderby Land, Antarctica. *Bulletin, Geological Society of America* 83, 3145–3150.

Mees, F. and Stoops, G. (1991) Mineralogical study of salt efflorescences on soils of the Jequetepeque Valley, North Peru. *Geoderma* 49, 255–272.

Mehta, P. K. (1983) Mechanism of sulphate attack on Portland cement concrete — another look. *Cement and Concrete Research*, 13, 401–406.

Melvin, J. L. (Ed.) (1991) *Evaporites, Petroleum and Mineral Resources*. Amsterdam: Elsevier.

Merrill, G. P. (1900) Sandstone disintegration through the formation of interstitial gypsum. *Science* NS 9, 850–851.

Miller, H. (1841) *The Old Red Sandstone*. Edinburgh: Nimmo.

Miller, W. R. and Mason, T. R. (1994) Erosional features of coastal beachrock and aeolianite outcrops in Natal and Zululand, South Africa. *Journal of Coastal Research* 10, 374–394.

Minty, E. J. (1965) Preliminary report on an investigation into the influence of several factors on the sodium sulphate soundness test for aggregate. *Australian Road Research* 2 (4), 49–52.

Minty, E. J. and Monk, K. (1966) Predicting the durability of rock. In: *Proceedings, 3rd Conference Australian Road Research Board 3*, 1316–1333.

Miotke, F. D. van. and Hodenberg, R. von (1980) Zur Salzsprengung und chemischen Verwitterung in den Darwin Mountains und den Dry Valleys, Victoria-Land, Antarktis. *Polarforschung* 50, 45–80.

Monger, H. C. and Daugherty, L. A. (1991) Pressure solution: possible mechanism for silicate grain dissolution in a petrocalcic horizon. *Journal Soil Science Society of America* 55, 1625–1629.

Moropoulou, T., Theoulakis, P., Kokkinops, C., Papsotiriou, D. and Daflos, E. (1992) The performance of some inorganic consolidants on the calcareous sandstone of the medieval city of Rhodes. In: *Proceedings, 7th International Congress on Deterioration and Conservation of Stone* (Eds J. Delgado Rodriguez, F. Henriques and F. Telmo Jeremias). Lisbon: Laboratorio Nacional de Enghenaria Civil, 1231–1242.

Moropoulou, A., Theoulakis, P., Bisbikon, K., Biscontin, G., Zendri, E., Bakolas, A. and Maravelaki, P. (1993) Behaviour of the bricks in the Corfu Venetian Fortress. In: *Conservation of Stone and Other Materials* (Ed. M.-J. Thiel). London: Spon, 402–410.

Mortensen, H. (1933) Die "Salzsprengung" und ihre Bedeutung für die regional klimatische Gliederung der Wüsten. *Petermanns Geographische Mitteilungen* 79, 130–135.

Moses, C. A. and Smith, B. J. (1994) Limestone weathering in the supra-tidal zone: an example from Mallorca. In: *Rock Weathering and Landform Evolution* (Eds D. A. Robinson and R. B. G. Williams). Chichester: Wiley, 433–452.

Mottershead, D. N. (1982) Coastal spray weathering of bedrock in the supratidal zone at East Prawle, South Devon. *Field Studies* 5, 663–684.

Mottershead, D. N. (1994) Spatial variations in intensity of alveolar weathering of a dated sandstone structure in a coastal environment, Weston-Super-Mare, U.K. In:

Rock Weathering and Landform Evolution (Eds D. A. Robinson and R. B. G. Williams). Chichester: Wiley, 151–174.

Mottershead, D. N. and Pye, K. (1994) Tafoni on coastal slopes, South Devon, UK. *Earth Surface Processes and Landforms* 19, 543–563.

Mueller, G. (1968) Genetic histories of nitrate deposits from Antarctica and Chile. *Nature* 219, 113–116.

Murata, K. J. (1977) Occurrence of bloedite and related minerals in marine shale of Diablo and Temblor Ranges, California. *Journal of Research, United States Geological Survey* 5, 637–642.

Murdock, L. J., Brook, K. M. and Jgwar, J. D. (1991) *Concrete Materials and Practice*. Arnold: London.

Mustoe, G. E. (1982) The origin of honeycomb weathering. *Bulletin of the Geological Society of America* 93, 108–115.

Mustoe, G. E. (1983) Cavernous weathering in the Capitol Reef Desert, Utah. *Earth Surface Processes and Landforms* 8, 517–526.

Mylona, S. (1996) Sulphur dioxide emissions in Europe 1880–1991 and their effect on sulphur concentrations and depositions. *Tellus* 48B, 662–689.

Netterberg, F. (1970) Occurrence and testing for deleterious salts in road construction materials with particular reference to calcretes. In: *Proceedings of the Symposium on Soils and Earth Structures in Arid Climates Adelaide*, 87–92.

Nichols, R. L. (1963) Geological features demonstrating aridity of McMurdo Sound area, Antarctica. *American Journal of Science* 261, 20–31.

Nichols, R. L. (1969) Geomorphology of Inglefield Land, North Greenland. *Meddelelser am Grønland* 188 (1).

Nord, A. G. (1992) Efflorescence salts on weathered building stone in Sweden. *Geologiska Föreningens I Stockholm Föhandlingar* 114, pt 4, 423–429.

Nord, A. G. and Ericsson, T. (1993) Chemical analysis of thin black layers on building stone. *Studies in Conservation* 38, 25–35.

O'Brien, P. F., Bell, E., Pavia Santamaria, S., Boyland, P. and Cooper, T. P. (1995) Role of mortars in the decay of granite. *The Science of the Total Environment* 167, 103–110.

Oberholster, R. E., Aardt Van, J. H. P. and Branot, M. P. (1983) Durability of cementitious systems. In: *Structure and Performance of Cements* (Ed. P. Barnes). London and New York: Applied Science Publishers, 55–63.

Obika, B., Freer-Hewish, R. J. and Fookes, P. G. (1989) Soluble salt damage to thin bituminous road and runway surfaces. *Quarterly Journal of Engineering Geology* 22, 59–73.

Orial, G. and Riboulet, G. (1993) Technique de nettoyage de la statuaire monumentale par désincrustation photonique. Réalisation d'un prototype mobile. In: *Conservation of Stone and Other Materials* (Ed. M.-J. Thiel). London: Spon, 542–549.

Orshan, G. (1986) The deserts of the Middle East. In: *Hot Deserts and Arid Shrublands* (Eds M. Evenari, I. Noy-Meir and D. W. Goodall). Amsterdam: Elsevier, 1–28.

Pantony, D. A. (1961) Sodium nitrate. In: *Comprehensive Treatise on Inorganic and Theoretical Chemistry*, Vol. II, Supplement II, The Alkali Metals, Part I (Ed. J. W. Mellor). London: Longmans, 1206–1258.

Peck, A. J. (1983) Response of groundwater to clearing in western Australia. In: *Papers, International Conference on Groundwater and Man* 327–335.

Pedro, G. (1957a) Mécanisme de la désagrégation du granite et de la lave de Volvic, sous l'influence des sels de cristallisation. *Comptes Rendu Academie des Sciences, Paris* 245, 333–335.

Pedro, G. (1957b) Nouvelles recherches sur l'influence des sels dans la désagrégetion des roches. *Comptes Rendu Academie des Sciences* 244, 2822–2824.

Peel, R. F. (1974) Insolation weathering; some measurements of diurnal temperature changes in exposed rocks in the Tibesti region, Central Sahara. *Zeitschrift für Geomorphologie, Supplementband* 21, 19–28.

Piqué, F, Dei, L. and Ferron, E. (1992) Physicochemical aspects of the deliquescence of calcium nitrate and its implications for wall painting conservation. *Studies in Conservation* 37, 217–227.

Pohl, E. R. and White, W. B. (1965) Sulphate minerals: the origin in the central Kentucky karst. *American Mineralogist* 50, 1461–1465.

Porto, J. C. (1977) Efflorescencias sulfatadas en formaciones terciarias y cuatarias de la Provincia de Tucuman. *Acta Geological Lilloana* 14, 261–276.

Prebble, M. M. (1967) Cavernous weathering in the Taylor Dry Valley, Victoria Land, Antarctica. *Nature* 216, 1194–1195.

Price, C. A. and Brimblecombe, P. (1994) Preventing salt damage in porous materials. In: *Preventive Conservation: Practice, Theory and Research*. International Institute for Conservation, London, pp. 90–93.

Prick, A. (1996) La désagrégation mécanique des roches par le gel et l'halocalstie. *Doctoral Thesis*, University of Liège, Belgium, 2 vols, 292 pp.

Pühringer, J. (1983) Salt disintegration. Salt migration and degradation by salt — a hypothesis. *Swedish Council for Building Research Document D15*, 159 pp.

Pye, K. and Mottershead, D. N. (1995) Honeycomb weathering of carboniferous sandstone in a sea wall of Weston-Super-Mare, UK. *Quarterly Journal of Engineering Geology* 28, 333–347.

Pye, K. and Schiavon, N. (1989) Cause of the attack on concrete, render and stone indicated by sulphur isotope ratios. *Nature* 343, 663–664.

Pye, K. and Sperling, C. H. B. (1983) Experimental investigation of silt formation by static breakage processes: the effect of temperature, moisture and salt on quartz dune sand and granitic regolith. *Sedimentology* 30, 49–62.

Reheis, M. C. and Kihl, R. (1995) Dust deposition in southern Nevada and California, 1984–1989: relations to climate, source area, and source lithology. *Journal of Geophysical Research* 100, 8893–8918.

Reheis, M. C., Goodmacher, J. C., Harden, J. W., McFadden L. D., Rockwell, T. K., Shroba, R. R., Sowers, J. M. and Taylor, E. M. (1995) Quaternary soils and dust deposition in southern Nevada and California. *Bulletin Geological Society of America* 107, 1003–1022.

Rhoades, J. D. (1990) Soil salinity — causes and controls. In: *Techniques for Desert Reclamation* (Ed. A. S. Goudie). Chichester: Wiley, 109–134.

Risacher, F. (1978) Le cadre géochimique des bassins à evaporites de Andes Boliviennes. *Cahiers ORSTOM, Series Géologique* 10, 37–48.

Robinson, D. A. and Williams, R. B. G. (1982) Salt weathering of rock specimens of varying shape. *Area* 14, 293–299.

Robinson, D. M. (1995) Concrete corrosion and slab heaving in a Sabkha environment: Long Beach–Newport Beach, California. *Environmental and Engineering Geoscience* 1, 35–40.

Rodriguez-Navarro, C. and Sebastian, E. (1996) Role of particulate matter from vehicle exhaust on porous building stones (limestone) sulphation. *The Science of the Total Environment* 187, 79–91.

Rögner, K. (1986) Temperature measurements of rock surfaces in hot deserts (Negev, Israel). *International Geomorphology Part II*. Chichester: Wiley, 1271–1286.

Rösch, H. and Schwartz, H. J. (1993) Damage to frescoes caused by sulphate-bearing salts: where does the sulphur come from? *Studies in Conservation* 38, 224–230.

Ross, K. D. and Butlin, R. N. (1989) Durability tests for building stone. *Building Research Establishment Report*, 8 pp.

Rossi-Manaresi, R. and Tucci, A. (1991) Pore structure and the disrupting or cementing effect of salt crystallization in various types of stone. *Studies in Conservation* 36, 53–58.

Rozanov, B. G., Targulian, V. and Orlov, D. S. (1991) Soils. In: *The Earth as Transformed by Human Action* (Ed. B. L. Turner). Cambridge: Cambridge University Press, 3203–14.

Rueffel, P. G. (1968) Development of the largest sodium sulphate deposit in Canada. *Mining and Metallurgical Bulletin* 61, 1217–1228.

Said, R. (1962) *The Geology of Egypt*. Amsterdam: Elsevier.

Salmon, L. G., Cass, G. R., Kozlowski, R., Heida, A., Spiker, E. C. and Bates, A. L. (1996) Air pollutant intrusion into the Wieliczka salt mine. *Environmental Science and Technology* 30, 872–880.

Schaffer, R. J. (1932) *The Weathering of Natural Building Stones*. Department of Scientific and Industrial Research Special Report 18. London: HMSO 149 pp.

Schemenauer, R. S. and Cereceda, P. (1992a) Monsoon cloudwater chemistry on the Arabian Peninsula. *Atmospheric Environment* 26A, 1583–1587.

Schemenauer, R. S. and Cereceda, P. (1992b) The quality of fog water collected for domestic and agricultural use in Chile. *Journal of Applied Meteorology* 31, 275–290.

Schiavon, N. (1993) Microfabrics of weathered granite in urban monuments. In: *Conservation of Stone and Other Materials*, Vol. 1. (Ed. M.-J. Thiel). London: Spon, 271–278.

Schiavon, N., Chiavari, G., Schiavon, G. and Fabbri, D. (1995) Nature and decay effects of urban soiling on granitic building stones. *The Science of the Total Environment* 167, 87–101.

Schneider, J. (1976) Biologic and inorganic factors in the destruction of limestone coasts. *Contributions to Sedimentology* 6, 1–112.

Schroeder, J. H. (1985) Eolian dust in the coastal desert of the Sudan: aggregates cemented by evaporites. *Journal of African Earth Science* 3, 370–386.

Scott, W. S. and Wylie, N. P. (1980) The environmental effects of snow dumping: a literature review. *Journal of Environmental Management* 10, 219–240.

Searl, A. and Rankin, S. (1993) A preliminary petrographic study of the Chilean nitrates. *Geological Magazine* 130, 319–333.

Seedsman, R. (1986) The behaviour of clay shales in water. *Canadian Geotechnical Journal* 23, 18–21.

Segerstorm, K. and Henriquez, H. (1964) Cavities or "tafoni" in rock faces of the Atacama Desert, Chile. *United States Geological Survey Professional Paper* 501C, 121–125.

Selby, M. J. and Wilson, A. T. (1971) The origin of the Labyrinth, Wright Valley, Antarctica. *Bulletin, Geological Society of America* 82, 471–476.

Selwitz, C. (1990) Deterioration of the Great Sphinx: an assessment of the literature. *Antiquity* 64, 853–859.

Shaw, G. E. (1991) Aerosol chemical components in Alaska air masses 2. Sea salt and marine product. *Journal of Geophysical Research* 96, 22369–22372.

Shehata, W. and Lotfi, H. (1993) Preconstruction solution for groundwater rise in Sabkha. *Bulletin of the International Association of Engineering Geology* 47, 145–150.

Simon, S. and Snethlage, R. (1996) Marble weathering in Europe—results of the Eurocare–Euromarble exposure programme 1992–4. In: *Proceedings, 8th International Congress on Deterioration and Conservation of Stone* (Ed. J. Riederer), Berlin: Ernst und Sohn, 159–166.

Sinclair, J. G. (1922) Temperatures of the soil and air in a desert. *Monthly Weather Review* 50, 142–144.

Skinner, B. J. (1966) Thermal expansion. *Geological Society of America Memoir* **97**, 75–96.

Slezak, L. A. and Last, W. A. (1985) Geology of sodium sulfate deposits of the northern Great Plains. In: *Proceedings Twentieth Forum on the Geology of Industrial Minerals Baltimore*, 105–115.

Smalley, I. J. (1966) The properties of glacial loess and the formation of loess deposits. *Journal of Sedimentary Petrology* **36**, 669–676.

Smalley, I. J. (1995) Making the material: the formation of silt-sized primary mineral particles for loess deposits. *Quaternary Science Reviews* **14**, 645–651.

Smalley, I. J. and Krinsley, D. H. (1978) Loess deposits associated with deserts. *Catena* **5**, 53–66.

Smalley, I. J. and Vita-Finzi, C. (1968) The formation of fine particles in sandy deserts and the nature of desert loess. *Journal of Sedimentary Petrology* **38**, 766–774.

Smirnioudi, V. N. and Siskos, P. A. (1992) Chemical composition of wet and dust deposition in Athens, Greece. *Atmospheric Environment* **26B**, 483–490.

Smith, B. J. and McAlister, J. J. (1986) Observations on the occurrence and origins of salt weathering phenomena near Lake Magadi, southern Kenya. *Zeitschrift für Geomorphologie* N.F. **30**, 445–460.

Smith, B. J. and McGreevy, J. P. (1983) A simulation study of salt weathering in hot deserts. *Geografiska Annaler* **65 A**, 127–133.

Smith, B. J., McGreevy, J. P. and Whalley, W. B. (1987) Silt production by weathering of a sandstone under hot arid conditions: an experimental study. *Journal of Arid Environments* **12**, 199–214.

Smith, S. E. (1986) An assessment of structural deterioration on ancient Egyptian monuments and tombs in Thebes. *Journal of Field Archaeology* **13**, 503–510.

Smoot, J. P. and Lowenstein, T. K. (1991) Depositional environments of non-marine evaporites. In: *Evaporites, Petroleum and Mineral Resources* (Ed. J. L. Melvin). Amsterdam: Elsevier, 189–347.

Soroka, I. (1993) *Concrete in Hot Environments*. London: Spon.

Sparks, B. W. (1949) The denudation chronology of the dip-slope of the South Downs. *Proceedings of the Geologists' Association of London* **60**, 165–215.

Sperling, C. H. B. and Cooke, R. U. (1985) Laboratory simulation of rock weathering by salt crystallization and hydration processes in hot, arid environments. *Earth Surface Processes and Landforms* **10**, 541–555.

St. John, D. A. (1982) An unusual case of ground water sulphate attack on concrete. *Cement and Concrete Research* **12**, 633–639.

Steiger, M. and Zeunert, A. (1996) Crystallization properties of salt mixtures: comparison of experimental results and model calculations. In: *Proceedings, 8th International Congress on Deterioration and Conservation of Stone* (Ed. J. Riederer). Berlin: Ernst und Sohn, 535–544.

Stephen, H. and Stephen, T. (1963) *Solubilities of Inorganic and Organic Compounds*, Vol. 1, Binary System, Part I. Oxford: Pergamon Press.

Stoertz, G. E. and Eriksen, G. E. (1974) *Geology of Salars in Northern Chile*. USGS Professional Paper **811**.

Strakhov, N. M. (1970) *Principles of Lithogenesis*, Vol. 3. Edinburgh: Oliver and Boyd.

Strzelczyk, A. B. (1981) Stone. In: *Microbial Biodeterioration* (Ed. A. H. Rose). London: Academic Press, 61–80.

Szabolcs, I. (1979) *Review of Research on Salt-affected Soils*. Paris: Unesco.

Szabolcs, I. (1994) State and perspectives on soil salinity in Europe. *European Society for Soil Conservation Newsletter* **3**, 17–24.

Taylor, R. K. and Smith, T. J. (1986) The engineering geology of clay minerals: swelling, shrinking and mudrock breakdown. *Clay Minerals* **21**, 235–260.

Tedrow, J. C. F. (1970) Soil investigations in Inglefield Land, Greenland. *Meddelelser am Grønland*, **188** (3).

Tedrow, J. C. F. (1977) *Soils of the Polar Landscape*. New Brunswick: Rutgers University Press.

Terzaghi, K. and Peck, R. B. (1948) *Soil Mechanics in Engineering Practice*. New York: Wiley.

Theoulakis, P. and Beloyannis, N. (1991) Stone deterioration problems caused by previous restorations at the Citadel of Lindos (Rhodes). In: *Science, Technology and European Cultural Heritage* (Eds N. S. Baer, C. Sabbioni and A. I. Sors). Oxford: Butterworth-Heinemann, 675–678.

Theoulakis, P. and Moropoulou, T. (1988) Mechanism of deterioration of the sandstone of the medieval city and castle of Rhodes. In: *Proceedings, 6th International Congress on the Deterioration and Conservation of Stone*, Torun: Nicolas Copernicus University Press Department, 86–96.

Thornton, T. G. and Nelson, K. H. (1956) Concentration of brines and deposition of salts from sea water under frigid conditions. *American Journal of Science* **254**, 229–238.

Tjia, H. D. (1985) Notching by abrasion on a limestone coast. *Zeitschrift für Geomorphologie* **29**, 367–72.

Tolba, M. K. and El-Kholy, O. A. (1992) *The World Environment 1972–1992*. London: Chapman and Hall.

Torfs, K., Van Grieken, R. and Cassar, J. (1996) Monitoring of environmental parameters to explain stone deterioration: church of Sta Marija Ta'Cwerra, Malta. In: *Proceedings, 8th International Congress on Deterioration and Conservation of Stone* (Ed. J. Riederer). Berlin: Ernst und Sohn, 264–271.

Traill, R. J. (1970) *A Catalogue of Canadian Minerals*. Geological Survey of Canada Paper **69**–45.

Trenhaile, A. S. (1987) *The Geomorphology of Rock Coasts*. Oxford: Clarendon Press.

Trenhaile, A. S. and Mercan, D. W. (1984) Frost weathering and the saturation of coastal rocks. *Earth Surface Processes and Landforms* **9**, 321–331.

Trenhaile, A. S. and Rudakas, P. A. (1981) Freeze–thaw and shore platform development in Gaspé, Québec. *Géographie Physique et Quaternaire* No. 2, 171–181.

Tricart, J. (1960) Expériences de désagrégation de roches granitiques par la cristallisation du sel marin. *Zeitschrift für Geomorphologie Supplementband* **1**, 239–240.

Tricart, J. (1962) Observations de Géomorphologie Littorale à Mamba Point (Monrovia, Libéria). *Erdkunde* **16**, 49–57.

Trudgill, S. T. (1985) Bioerosion of intertidal limestone, Co. Clare, Eire — 3: zonation, process and form. *Marine Geology* **74**, 111–121.

Tuncoku, S. S., Calner-Saltik, E. N. and Böke, H. (1993) Definition of the materials and related problems of a XIIIth century Anatolian Seljuk 'Mescid': a case study in Konya City. In: *Conservation of Stone and Other Materials* (Ed. M.-J. Thiel). London: Spon, 368–375.

Turner, E. (1833) Report on a lecture on the chemistry of geology. *London and Edinburgh Philosophical Magazine, Journal of Science* **3**, 21.

Unesco (1964) *Preservation of the Monument of Mohenjo-Daro, Pakistan*. Paris: Unesco, 63.

van Aardt, J. H. P. and Visser, S. (1975) Thausamite formation: a cause of deterioration of Portland cement and related substances in the presence of sulphates. *Cement and Concrete Research* **5**, 225–237.

Van Hees, R., Van der Klugt, L. J. A. R., Naldini, S., Van Baalen, K., Mascarenhas Mateus, J., Binda, L., Baronio, G., Francke, L. and Schumann, I. (1996) The masonry damage diagnostic system. In: *Proceedings, 8th International Congress on Deterioration and Conservation of Stone* (Ed. J. Riederer), Berlin: Ernst und Sohn, 1737–1740.

van Lier, J. A., de Bruyn, P. L. and Overbeck, J. Th. G. (1960) The solubility of quartz. *Journal of Physical Chemistry* 64, 1675–1682.

Vassie, P. (1984) Reinforcement corrosion and the durability of concrete bridges. In: *Proceedings, Institution of Civil Engineers* 76, 713–723.

Vendrell-Saz, M., Krumbein, W. E., Urzì, C. and Garcia-Vallès, M. (1996) Are patinas of Mediterranean monuments really related to the rock substrate? In: *Proceedings, 8th International Congress on Deterioration and Conservation of Stone* (Ed. J. Riederer). Berlin: Ernst und Sohn, 609–623.

Verbeck, G. J. and Klieger, P. (1957) Studies of "salt" scaling on concrete. *Highways Research Board Bulletin* 150, 1–13.

Vergès-Belmin, V. (1995) Pseudomorphism of gypsum after calcite. A new textural feature accounting for the marble sulphation mechanism. *Atmospheric Environment* 28, 295–304.

Viles, H. A. (1990) The early stages of building stone decay in an urban environment. *Atmospheric Environment* 24A, 229–232.

Viles, H. A. (1993a) Observations and explanations of stone decay in Oxford, UK. In: *Conservation of Stone and Other Materials* (Ed. M.-J. Thiel). London: Spon, 115–120.

Viles, H. A. (1993b) The environmental sensitivity of blistering of limestone walls in Oxford, England: a preliminary study. In: *Landscape Sensitivity* (Eds D. S. G. Thomas and R. J. Allison). Chichester: Wiley, 309–326.

Villegas, R. and Vale, J. F. (1992) Evaluation of the behaviour of water repellent treatments for stone. In: *Proceedings, 7th International Congress on Deterioration and Conservation of Stone* (Eds J. D. Rodrigues, F. Henriques and F. Telmo Jeremias). Lisbon: Laboratorio Nacional de Enghenaria Civil, 1253–1262.

Villegas Sánchez, R., Martín Gárcia, L., Vale Parapar, J. F., Bello Lopéz, M. A. and Alcalde Moreno, M. (1996) Characterization and conservation of the stone used in the Cathedral of Almeria (Spain). In: *Proceedings, 8th International Congress on Deterioration and Conservation of Stone* (Ed. J. Riederer). Berlin: Ernst und Sohn, 89–99.

Vitina, I., Iguane, S., Krage, L. and Baumanis, O. (1996) Problems of soluble salts in the monuments of Latvia. In: *Proceedings, 8th International Congress on Deterioration and Conservation of Stone* (Ed. J. Riederer). Berlin: Ernst und Sohn, 477–480.

Vleugels, G., Roekens, E., Van Put, A., Araujo, F., Fobe, B., Van Grieken, R., Mesquita e Carmo, A., Azevedo Alves, L. and Aires–Barros, L. (1992) Analytical study of the weathering of the Jeronimos Monastery in Lisbon. *The Science of the Total Environment* 120, 225–243.

Voûte, C. (1962) Some geological aspects of the conservation project for the Philae temples in the Aswan area. *Geologische Rundschau* 52, 665–675.

Wali, A. M. A. (1991) Sabkhas of the Bitter-Lakes, Egypt: composition and origin. *Carbonate and Evaporites* 6, 225–238.

Wallace, R. C. (1916) The corrosive action of certain brines in Manitoba. *Geological Magazine* 3, New Decade Series IV, 31–32.

Walter, H. (1937) Die ökologisches Verhältnisse in der Namib Neblewüste (Süduestafrika) unter Answertung der Aufzeichnungen des Dr. G. Boss (Swakopmund). *Jahrbuch für Wissenschaftliche Botanik* 84, 88–222.

Walther, J. (1912) *Das Gesetz der Wüstenbildung in Gegenwart und Vorzeit*. Berlin: Quelle and Mayer. 342 pp.

Wand, U. (1995) Salt efflorescences. *Petermanns Geographische Mitteilungen Ergänzungsheft* **289**, 201–204.

Warburg, M. R. (1964) Observations on microclimate in habitats of some desert vipers in the Negev, Arava and Dead Sea Region. *Vie et Milieu* **15**, 1017–1041.

Warn, G. F. and Cox, W. H. (1951) A sedimentological study of dust storms in the vicinity of Lubbock, Texas. *American Journal of Science* **249**, 553–568.

Warren, J. K (1989) *Evaporite Sedimentology*. Englewood Cliffs: Prentice Hall.

Washburn, A. L. (1969) Weathering, frost action and patterned ground in the Mesters Vig district, North East Greenland. *Meddeleser am Grønland* **176** (4).

Watson, A. (1982) The origin, nature, and distribution of gypsum crusts. *Unpublished D. Phil Dissertation*, University of Oxford, 669 pp.

Watson, A. (1983) Gypsum crusts. In: *Chemical Sediments and Geomorphology* (Eds A. S. Goudie and K. Pye). London: Academic Press, pp 133–161.

Watts, N. L. (1977) Pseudo-anticlines and other structures in some calcretes of Botswana and South Africa. *Earth Surface Processes* **2**, 63–74.

Watts, S. H. (1979) Some observations on rock weathering, Cumberland Peninsula, Baffin Island. *Canadian Journal of Earth Sciences* **16**, 997–983.

Watts, S. H. (1981) Near coastal and incipient weathering features in the Cape Herschel–Alexandra Fjord area, District of Franklin. *Geological Survey of Canada, Paper*, 81–1A, 389–394.

Weiss, G. (1992) Die Eis und Salzkristallisation im Porenraum von Sandsteinen und ihre Auswirkungen auf das Gefüge untger besonderer Berücksichtigung gesteins-spezfischer Parameter. *Münchner Geowissenschaftliche Abhandlungen, Reihe B, Allgemeine und Angewandte Geologie* **9**, 112 pp.

Wellman, H. W. and Wilson, A. T. (1965) Salt weathering: a neglected geological erosive agent in coastal and arid environments. *Nature* **205**, 1097–1098.

Wells, A. T. (1976) Salt—Nothern Territory. In: *Industrial Minerals and Rocks* (Ed. C. L. Knight). Australian Institute of Mining and Metallurgy, pp. 341–344.

Wells, R. C. (1923) Sodium sulphate: it sources and uses. *United States Geological Survey, Bulletin* **717**, 43.

Wessman, L. (1996) Studies of salt-frost attack on natural stone. In: *Proceedings, 8th International Congress on Deterioration and Conservation of Stone* (Ed. J. Riederer), Berlin: Ernst und Sohn, 563–571.

Whalley, W. B., Smith, B. J. and Magee, R. (1992) Effects of particulate air pollutants on materials: investigation of surface crust formation. In: *Stone Cleaning* (Ed. R. G. M. Webster). London: Donhead Publishing, 27–34.

White, W. B. (1976) Cave minerals and speleothems. In *The Science of Speleology* (Ed. T. D. Ford and C. H. D. Cullingford). London: Academic Press, 267–327.

White, W. B. (1988) *Geomorphology and Hydrology of Karst Terrains*. New York: Oxford University Press, 464 pp.

Whittig, L. D., Deyo, A. E. and Tanji, K. K. (1982) Evaporite mineral species in Mancos Shale and salt efflorescence, Upper Colorado river basin. *Soil Science Society of America Journal* **46**, 645–651.

Williams, C. B. (1923) A short bio-climate study in the Egyptian desert. *Ministry of Agriculture, Egypt, Technical and Scientific Service Bulletin* **29**, 1–19.

Williams, R. B. G. and Robinson, D. A. (1981) Weathering of sandstone by the combined action of frost and salt. *Earth Surface Processes and Landforms* **6**, 1–9.

Williams, R. B. G. and Robinson, D. A. (1991) Frost weathering of rocks in the presence of salts—a review. *Permafrost and Periglacial Processes* **2**, 347–353.

Wilson, A. T. (1979) Geochemical problems of the Antarctic dry areas. *Nature* 280, 205–208.

Winkler, E. M. (1975) *Stone: Properties, Durability in Man's Environment* (2nd edn). Vienna: Springer Verlag, 230 pp.

Winkler, E. M. (1979) Role of salts in development of granitic tafoni, south Australia: a discussion. *Journal of Geology* 87, 119–120.

Winkler, E. M. (1981) Salt weathering by sodium chloride in the Saudi Arabian Desert. *American Journal of Science* 281, 1244–1245.

Winkler, E. M. and Singer, P. C. (1972). Crystallization pressure of salts in stone and concrete. *Geological Society of America Bulletin* 83, 3509–3514.

Winkler, E. M. and Wilhelm, E. J. (1970) Saltburst by hydration pressures in architectual stone in urban atmosphere. *Bulletin Geological Society of America* 81, 567–572.

Wittenburg, C., Wendler, E. and Steiger, M. (1996) Terracotta at Schloss Schwerh, different desalination treatments for the application of stone consolidating agents. In: *Proceedings, 8th International Congress on Deterioration and Conservation of Stone* (Ed. J. Riederer). Berlin: Ernst und Sohn, 1717–1726.

Woodward, H. P. (1897) The dry lakes of Western Australia. *Geological Magazine* N.S., Decade IV, 4, 363–366.

World Resources Institute (1987) *World Resources 1987–1988*. Oxford: Oxford University Press.

Yaalon, D. H. (1965) Source and sedimentary history of the loess in the Beer Sheva Basin, Israel. In: *INQUA 7th Congress Abstracts*, 514.

Yaalon, D. H. and Ginzbourg, D. (1966) Sedimentary characteristics and climatic analysis of easterly dust storms in the Negev (Israel). *Sedimentology* 60, 315–322.

Young, A. R. M. (1987) Salt as an agent in the development of cavernous weathering. *Geology* 15, 962–966.

Young, R. and Young, A. (1992) *Sandstone Landforms*. Berlin: Springer Verlag.

Zangvil, A. (1996) Six years of dew observations in the Negev Desert, Israel. *Journal of Arid Environments* 32, 361–371.

Zehnder, K. (1993) New aspects of decay caused by crystallization of gypsum. In: *Conservation of Stone and Other Materials* (Ed. M.-J. Thiel). London: Spon, 107–114.

Zehnder, K (1996) Gypsum efflorescence in the zone of rising damp. Monitoring of slow decay processes caused by crystallizing salts on wall paintings. In: *Proceedings, 8th International Congress on Deterioration and Conservation of Stone* (Ed. J. Riederer). Berlin: Ernst und Sohn, 1669–1678.

Zehnder, K. and Arnold, A. (1984) Stone damage due to formate salts. *Studies in Conservation* 29, 32–34.

Zeuner, F. G. (1949) Frost soils on Mount Kenya. *Journal of Soil Science* 1, 20–30.

Zezza, F. and Macrì, F. (1995) Marine aerosol and stone decay. *The Science of the Total Environment* 167, 123–143.

Zheng, M., Tang, J., Lui, J. and Zhang, F. (1993) Chinese saline lakes. *Hydrobiologia* 267, 13–21.

Index